DATE DUE

DEMCO 38-296

Analyzing Field Measurements
AIR CONDITIONING
& HEATING

Analyzing Field Measurements
AIR CONDITIONING
& HEATING

Robert S. Curl, P.E.
Energy Engineer

Published by
THE FAIRMONT PRESS, INC.
700 Indian Trail
Lilburn, GA 30247

TH 7687.5 .C87 1997

Curl, Robert S.

Analyzing field measurements

...ublication Data

...conditioning & heating / Robert S.

...oling load--Measurement.
3. Heating load--Measurement. 4. Heating. I. Title.
TH7687.5.C87   1997   697--dc21   96-54291
                                                      CIP

Analyzing Field Measurements : Air Conditioning & Heating By Robert S. Curl, Second Edition.

Published by The Fairmont Press, Inc.
700 Indian Trail
Lilburn, GA 30247

Printed in the United States of America

10 9 8 7 6 5 4 3 2 1

0-88173-254-0 FP

0-13-760133-6 PH

Distributed by Prentice Hall PTR
Prentice-Hall, Inc.
A Simon & Schuster Company
Upper Saddle River, NJ 07458

Prentice-Hall International (UK) Limited, London
Prentice-Hall of Australia Pty. Limited, Sydney
Prentice-Hall Canada Inc., Toronto
Prentice-Hall Hispanoamericana, S.A., Mexico
Prentice-Hall of India Private Limited, New Delhi
Prentice-Hall of Japan, Inc., Tokyo
Simon & Schuster Asia Pte. Ltd., Singapore
Editora Prentice-Hall do Brasil, Ltda., Rio de Janeiro

Table of Contents

Introduction

This manual is intended to assist engineers and technicians, with a background of use of temperature, pressure, humidity, air flow and flue gas instruments, to interpret their readings of field measurements. This determination will indicate the deficiencies of the system tested and methods of correction. Improvement of operation nearly always results in reduction in utility costs.

The systems should be corrected to operate at a good condition before considering the purchases of new equipment or management control. Analysis of utility costs will indicate feasibility of the expenditure.

Most of the suggestions are based on the author's testing experiences from 1938 to 1991 as a Designer, Energy Analyst, Contractor, Serviceman and Consultant on heating, air conditioning, boiler plants, refrigeration, electric power and light, plumbing, piping, cogeneration, industrial and chemical process in over a billion dollars of buildings.

The author wishes to express his appreciation to many of the manufacturers of related equipment for some of the information and diagrams in this manual.

Case studies listed have been performed by the author. Many of the recommendations have been implemented with equivalent or better operational savings.

ANALYSIS OF FIELD MEASUREMENTS OF AIR CONDITIONING TO REDUCE ENERGY COSTS

The lowest cost method of reducing utility bills is to improve the operation of existing equipment. After the first five years, nearly all systems slip in efficiency. Like an automobile, the system needs an analysis, adjustment and retrofit.

This manual describes how to accomplish the programs to correct the operation of air conditioning equipment and accessory ducts, piping, machinery and other devices.

In many cases, the purchase of new equipment may not be necessary, or considerably less than replacement. Even if new machinery or management controls are more efficient, repair or retrofit can return investment in a much shorter time period.

Although many manufacturers supply good operation and maintenance manuals, there is limited information on how to provide for individual parts of different make to work together.

Learn how to use data obtained from field measurements to determine what needs to be done to get operating costs down.

The age of equipment does not determine efficiency or necessity for replacement. Some equipment over 30 years old, properly maintained, can be as good or better than new. Today's machinery is mostly lighter, faster, with shorter life.

Examples of Field Test Cost Savings

Residence – Reduced monthly electric bill from $6,100 to $2,900 per month.

Office Building – Saved $.65 per square foot adjusting cooling and lighting during cleaning.

Office Building – Saved $300,000 per year by adjusting cooling plant.

Hospital – Reduced energy bill 35% by correcting controls.

Country Clubhouse – Twenty-three years old, electric bill at $2.75 per square foot after retrofit. Comparative clubhouse, two years old: $4.65 per square foot.

School – Saved $42,000 per year adjusting power plant and piping distribution.

Electric Plant Clean Room – Saved $72,000 per year adjusting and correcting air conditioning system.

Hotel – Eliminated mildew damage in empty guest rooms.

Chapter 1

Initial Field Cooling Inspection

A. <u>AIR</u>

Since the use of air provides the means to make an occupied space comfortable, we need to know how to handle it, such as:

1. Temperature and Humidity Conditions
   a. Room
   b. Supply
   c. Internal Effects
   d. External Effects

2. How Air Is Supplied To the Space
   a. Ceiling outlets
   b. Side wall outlets
   c. Floor outlets
   d. Height above occupant level

3. Physical Activity of Occupants
   a. Seated
   b. Standing
   c. Constantly Moving
   d. Operating Machinery

4. Auxiliary Means of Air Motion
   a. Ceiling Fans

5. Air Supply Quantity

6. Description

In southern and coastal climates, the humidity is more important than the temperature. It takes nearly twice the cooling capacity to remove the moisture than to reduce the temperature.

(1a) The author's experience indicates that about 75° dry bulb and 60% relative humidity or less is the condition most people feel comfortable. The dew point at that level is 60°. This means that any surface at a temperature below 60° will get wet.

(1b) If any moisture is generated in the conditioned space (people give off at least 40% of their heat as perspiration), the supply air must be able to absorb it. The supply air should be below 60° to soak up much moisture.

The lower the supply air temperature the more moisture it will pick up. There is a limit, because air outlets may drip water and drafts may occur. An alternative is to over cool the air and reheat, but this is very energy inefficient.

(1c) Heat radiated to occupants from office equipment, lights, and machinery may require lower room temperatures or spot cooling.

(1d) Radiation from warm windows, walls, ceilings and even floors can upset comfort conditions and may require temperature changes in certain areas, some periods of the day.

(2a) Ceiling air outlets induce about 40% room air so the supply air fed into the area is warmer than that in the duct. Adjustable directional outlets may be better than fixed pattern so that air can be directed to usable areas. The problem with adjustability is that unauthorized persons sometimes move the blades.

(2b) Side wall grilles also induce air and should be adjustable. There must be enough velocity of air to keep it above people's heads before it mixes with the room air.

(2c) If air is supplied from floor, window sill or cabinet cooler, it also must have enough velocity to reach the whole area before falling to the floor.

(2d) One of the best methods of reducing cooling load in rooms with ceilings above 10 feet is to supply the air at 10 feet and let the warm air above lay dormant.

[Experience] In a church with a 30ft. ceiling, with air supplied at 11 ft. and a room temperature of 75°, tests showed the air at 14 ft. was 80° and at 16 ft. was 90°.

(3a) Seated people generate less heat than those active. As a result they may be comfortable at higher temperatures. Also they are susceptible to air motion, either too much or too little.

(3b) Standing still can result in the same conditions as seated except there is more fatigue. This has a tendency to require cooler air.

(3c, 3d) Activity increases body heat generation requiring lower temperature or higher air motion.

(4a) ASHRAE Research has indicated that the comfort level may increase as much as 6° F. with air velocities of 300 ft. per min. Ceiling fans do a good job of this. Again keep the fans low enough so as not to disturb the air above.

(5) For some reason, nearly every system the author has tested is low on air CFM. Either the belts are slipping, if the fan is forward curve blade they are dirty or rust, or the system has high static pressure. Low air volume means lower supply temperature to try to meet cooling demand.

INITIAL FIELD AIR CONDITIONING COOLING INSPECTION

Basis: Outside 85° – 95° / Inside 74° – 76°

B. SPACE CONDITIONS

1. Measure room temperature.

2. Read thermostat setting.

3. Measure supply air temperature at ceiling outlet or wall grille.

4. If room temperature is 2° or more above thermostat setting, supply air temperature should be below 60°. If above 60°, go to Step 5.

5a. Set thermostat at least 5° below room temperature. Measure supply air temperature. Air temperature should go down. If temperature drops below 54° there may be restricted air flow.

5b. Set thermostat 5° above room temperature. Measure air temperature. It should not go up to much over 80°. Above 80° indicates reheat or high outside air volume.

6a. Locate return grille. If room has lay-in ceiling and space above is used for return, measure air temperature above ceiling in area closest to air handler. Return below room air temperature indicates duct leaks.

6b. Fiberglass duct leaks – look for loose tape at joints and at tap in connections to main trunks.

Sheet metal – look for dirt streaks at joints.

7. If room is larger than 100 sq. ft., take a number of room temperature readings, at least every 500 sq. ft. If variation temperature is more than 2°, look for reason:

    a. Location of air outlet.

    b. Auxiliary heat source – office equipment, lighting, solar heat from windows or walls, concentration of people.

    c. If warm spot is more than 20' from air outlet, air velocity may not be high enough to carry air that distance. Outlet should be adjusted or air volume increased.

    d. Visible air motion or drafts. Air outlets should be adjusted to carry air overhead.

8. If ceiling height is above 12', and air outlets at this level or higher, warm air near ceiling is being picked up and mixing with cooled air. Examine the possibilities of lowering outlets.

Measure air temperature 6" from ceiling, 5' above floors, 6" above floor. Adjust outlets for even temperature floor to 6' above.

9. If ceiling fans exist they should not be higher than 12". If higher, they also pick up warm air. Can they be lowered?

C. AIR HANDLING EQUIPMENT CONDITIONS

    1. Measure supply air temperature in duct leaving unit.

    2. Compare temperature with air outlet reading. Temperature rise of more than 2° indicates insulation problems.

3. Determine type of system:
   a. Single zone
   b. Multizone
   c. Reheat
   d. Medium pressure mixing box
   e. Variable air volume
   f. Induction

4. Measure air temperature in return air duct.

5. Measure air temperature in outside air duct.

6. Measure temperature of mixed air entering unit.

7. Calculate percentage of outside air from temperatures 4, 5, 6. M.A. temperature – R.A. temperature/O.A. temperature – R.A. temperature X 100

8. Determine if this much outside air is necessary (20 CFM per person).

9. Examine the cooling coil:
   a. Are fins clean?
   b. If end of coil is exposed, feel return bends, top to bottom, for same level of cold.
   c. How many rows of tubes. Four (4) rows should be minimum, six (6) preferred.

10. If unit has access door, stop fan.
    a. Is there open space around coil that air could bypass?
    b. Feel fan blades. If dirty or rusted, air volume is reduced.

11. Check fan belts.

12. When running, measure amps and volts on fan motor. Check with nameplate rating. Low amps may indicate low fan capacity. Low air flow requires lower supply air temperature to produce enough cooling, consequently more kw to run systems.

13. Inspect filters for cleanliness.

14. Inspect unit housing for air leaks.

15. Measure air static pressure entering and leaving unit.
    a. Most air conditioning systems with extended ducts require a supply static pressure of about 1" W.G. Mixing box and VAV require up to 4" W.G. High static pressures indicate

restrictions in ducts. This is usually caused by dampers or turning vanes. Very low static may indicate duct leaks.

D. REFRIGERATION EQUIPMENT CONDITIONS

   1. Air Cooled Units

      a. Measure air temperature entering condenser coil and leaving condenser fan. About 20° rise is normal. Compare with outside temperature. Does discharge air recirculate?

      b. Measure unit amps and volts and compare with nameplate.

      c. If there is more than one condenser fan and they are not all running, unit is not at full load. Amps should not be up to full load.

   2. Water cooled units

      a. Measure water temperature entering and leaving condenser. 10° rise is normal. Lower temperature rise indicates unit is not fully loaded. Higher rise indicates low water flow, maybe pump problems.

   3. Condenser condition

      Air cooled –
      Is condenser coil clean?   _____
         Is there any obstruction on air
         entering or leaving side?   _____

      Water cooled –
         Have water side tubes been
         cleaned in last 2 years?   _____

   4. Location of condensing unit –

      If condensing unit is above air handler, refrigerant piping must be sized for distance and height. If refrigerant gauges are installed, look at readings. Compare with pressure temperature charts for the refrigerant used. Suction pressure should be equivalent to 40° or 50°, discharge pressure 95° to 115° when fully loaded. If there is no suction line trap, coil may be partially full of oil. Feel lower tubes of coil. They may be warmer.

E. CHILLED WATER SYSTEMS

   1. Measure water pressure at highest point of system. This should be at least 10 psi to make sure all equipment has water flowing

through it. If under 10 psi, locate makeup water valve, expansion tank, relief valve, and adjust. If relief valve opens at 12psi, replace with 30 psi valve.

2. Measure chilled water temperature entering and leaving chiller. Leaving temperature should be 42° to 45° for dehumidification. Temperature rise outlet to inlet should be from 8° to 12°. Higher rise indicates too low water flow. Low rise may indicate mineral coated water tubes in chiller.

3. Distribution Piping Analysis:

   Most chilled water systems supply multiple air handling units. Temperature control on each unit usually is a two position or three way valve.

   If two position is used, some means of bypassing water from the supply to return piping is necessary to prevent overloading pumps.

   If three-way valves are used full circulation is available all the time. Check the manufacturers capacity curves on the pumps for the most efficient operation.

   On three-way valves, it may be possible to close the bypass, then provide a control at the pump to put part of the supply water back into the return at the chiller. This reduces the pressure drop in the piping and the power for the pump motor.

   CAUTION: The chiller requires a minimum water flow to prevent freeze-up. Measure supply pipe surface temperatures and water pressures at various parts of the system to determine flow balance. Depending on pump operation, one method of using less chilled water at low loads is to close the bypass on all valves except a few at the end of the piping. Pump discharge pressure control may stop and start a second pump with a time delay.

F. AIR DISTRIBUTION SYSTEMS

1. High Supply Air Static Pressure
   a. Measure Static pressure entering and leaving
      1. Turning Vanes
      2. Fire Dampers
      3. Manual Dampers
      4. Mixing Boxes
      5. VAV Boxes

6. Behind air outlet dampers
7. Plug hole in duct and cover with duct tape or plug when finished testing

b. Check damper linkage, correct
   1. Is arm loose?
   2. Is damper fully open or closed at end of arm stroke?
   3. On mixing dampers, is one fully open when other fully closed?
   4. Does motor operate damper at full stroke?
   5. On fire dampers, has link stretched or loosened? Is damper fully open?
   6. Do dampers move easily? Lubricate if necessary.

c. Problems
   1. Inside duct insulation comes loose and stops at turning vanes and dampers.
   2. Dampers partially or totally closed. Reset dampers.
   3. Sharp duct turns close to supply fan outlet cause interference.
   4. Dirty reheat coils.
   5. Complicated duct elbows. (Check static pressure entering and leaving assembly. Compare with duct static chart.)
      [EXPERIENCE] Two 90° elbows together 20 ft from fan on 4 inch static system has 3 inch pressure drop. Changed to three 45° elbows, static now ½ inch.
   6. In double duct system, balance static pressure in both ducts under heating and cooling conditions.
   7. Low air supply static pressure.

a. If return air temperature is low, look for duct leaks.

b. Check fan rotation, speed, amps, belts and condition and compare with fan curves. Note: some forward curve fans may require a discharge aeliron.

[EXPERIENCE] An 8,000 CFM A/C unit with two forward curve fans was operating at 4,000 CFM at full electric load. Tee duct connection 4 feet from fan discharge interrupted air pattern. Aeliron installed in fan scroll discharge reduced the opening and corrected the capacity.

    c. If system is blow through, look for air bypassing coils or filters.

3. Air Quantity

    a. If air quantity supplied from air outlets is below total air supply

        1. Look for duct leaks.

            a. Trunk ducts.

            b. At take off branch connections.

            c. At connections from duct to air outlets.

            [EXPERIENCE] In testing 100 houses, the connection between the duct and drywall ceiling air outlet leaked in 56. SMACCNA calls for a sheet metal collar between trunk and air outlet.

    b. Check fan operation if air quantity is low.

        1. See 2b.

            [EXPERIENCE] A clean room operating at $70° - 30\%$ RH operated at 9,000 CFM at $0°$ with 228 tons. Increasing CFM to 28,000 at $56°$ reduced load to 76 tons.

        2. Check inlet condition of fan. If provided with inlet vanes, they should not be perpendicular to inlet. Stop at $15°$.

        3. If fans are axial, check MFR for distance of inlet and outlet straight duct required.

        4. Check abrupt turns in ducts close to fan and unit which may change air pattern entering or leaving fan. See EXPERIENCE 2b, lb(5) .

        5. Check for dirty or rusty blower blades, especially forward curve. Clean carefully.

        6. Check for wheel out of balance.

4. Air Patterns

    a. Side Wall Grilles.

        1. If air does not reach opposite wall adjust front blades, 1/3 bottom turned up, 1/4 top turned down. If front blades are vertical, adjust rear blades similarly. If rear blades are tied together set all up about half way from horizontal.

        2.   In winter to get heat to floor tilt blades downward.

        3.   Adjust vertical blades for room coverage.

  b.   Ceiling outlets.

        1.   If the outlet is non adjustable and is near a wall, remove and close the outside supply suction near wall, with tape, on the air entering side.

        2.   If the outlet is on a high ceiling and air doesn't get to the floor, remove the center section of outlet or provide a 1/2" angle around the outside top edge.

        3.   Inspect linear outlets for damper position. Some provide directional air flow.

  c.   Return air outlets.

        1.   Measure air quantity at return air grilles and at A/C unit to determine if return duct leaks air. Use a box the size of the grille at least 12" deep to get a good air reading.

        2.   Return air grille location is immaterial for cooling but must be near floor for heating. If high, use supply grille adjustment to force warm air to floor.

  d.   Air noise.

        1.   Air noise in air outlets is usually caused by dampers in back, or if the outlet is too small for the air quantity, try rotating the damper 90°.

  e.   Unbalanced air supply.

        1.   If the air supply varies between outlets on a main duct, examine the branch connection for a splitter damper or baffle. Air has a tendency to go in a straight line and pile up at the end of the duct.

G. CONTROLS

  1. Basic – Single Zone

    a.   Tests under air handler conditions should indicate if thermostat works.

    b.   On direct expansion systems, thermostat may start and stop compressor.

c. If there is a solenoid valve on the liquid line to the coil, the thermostat may operate valve and the compressor may cycle on a low pressure switch. If there is more than one air handler on a single compressor system, this is probably the control, with the compressor also internally unloaded by refrigerant pressure.

2. Other systems.

a. On nearly all other systems, the control of the compressor is by a thermostat in the air leaving the cooling coil. This thermostat may control solenoid valves on liquid lines to coil or face and bypass dampers on coil, The latter control provides better dehumidification. An end switch on damper motor should control compressor if face damper is almost closed.

3. Mixing damper systems. (Multizone)

a. Thermostat gradually moves dampers on zone ducts to proportion the amount of cooled and uncooled air to provide the correct supply temperature to the zone.

b. Field Observations: If some zones are cooled and others are not, examine the mixing dampers and damper motor with someone moving the thermostat setting. Dampers may leak even when closed. Look at damper arm. Does the damper motor move the dampers all the way open and shut?

c. Damper operation of zones and also outside and return air dampers can be checked with a thermometer. Compare hot deck, cold deck and zone supply air temperature when damper is fully closed in both positions. On air entering side check mixed air temperature with return air with outside damper closed. Compare mixed with outside air with return damper closed.

4. Chilled water systems.

a. Compressors are usually controlled by a thermostat in the return water line entering the chiller. This eliminates short cycling if thermostat was in the supply line. The control thermostat therefore is set 8° to 12° above the supply water temperature.

b.   Control at air handlers is by thermostat operating valves or dampers similar to direct expansion. Valves may be three-way bypassing water around the coil so as to keep the water flow in the system constant.

c.   There is a minimum water flow permitted for each chiller to prevent freeze up and equipment damage.

5. Thermostat location.

a.   Is thermostat located where it measures the average room temperature?

b.   Is there any heat producing source near thermostat?

c.   Does sun shine on thermostat?

d.   Is thermostat set by one person or do many people keep readjusting it on request or whim? Is there a locking cover on thermostat?

6. Heating.

a.   Hot water systems operate similar to chilled water with two position or three way valves. Water temperature should be kept as low as possible to prevent over- heating, especially on reheat systems.

b.   Electric heaters should be turned on in steps if larger than 5KW. Heater element stays warm as long as 20 minutes after shut off.

H. COOLING TOWERS

1. Measure entering and leaving dry bulb and wet bulb air temperatures.

2. Measure fan air quantity.

3. Measure entering and leaving water temperature.

4. Examine tower fill for surface coating such as slime, algae, minerals. Inspect drain pan for cleanliness.

5. Observe level of water in basin.

6. Observe pattern of water entering return pipe.

7. Inspect water distribution.

8. Compare entering air with outside air temperature.

9. Is fan single or two speed?

10. What is the control on fan?

11. Check water quantity overflowing to drain.

12. Field Observations

   (1) Wet bulb temperature leaving tower should be 3° to 8° above entering WB.

   (2) Check fan air quantity with manufacturers rating. too low reduces capacity. Too high blows water with air to form a plume. This loss of water is expensive.

   (3) Most towers are based on a 10° drop. If lower, items 2, 4, 6, 7, & 10 may be the cause.

   (4) Cooling of water depends on having the greatest wet surface exposed to moving air for absorption. Condition of fill seriously affects this.

   (5) Most towers require 4" or more water in basin or drain pan. This is to reduce pickup of dirt and air into the leaving water line. Air in line upsets pump operation.

   (6) Vortexing of water at outlet picks up air also.

   (7) Water must be distributed evenly over the tower to get the full capacity.

   (8) If entering air is warmer than outside air look for possible recirculation of discharge into intake due to walls, obstructions or wind.

   (9) Multi-speed or multiple fans do a better water cooling job than cycling fan on and off.

   (10) Overflow is normally 1% of total flow, Higher, causes expensive water loss. Check chemical treatment for overflow requirements.

I. SPECIFIC APPLICATIONS OF FIELD MEASUREMENT

   1. Department Stores.

      a. Entrance door infiltration:

         1. Measure air velocity and dimensions of opening. Determine CFM.

         2. If air direction is out, look for excess outside air supply.

         3. If air direction is in:

     a. Measure air direction into elevators. If strong, go to penthouse and look for ventilators or exhaust fans.

     b. Check kitchen for exhaust without makeup air.

     c. Check exhaust in storage, work areas, auditoriums and toilets.

b. Air doors.

    1. Measure direction of air in or out near floor and at top of opening. Outside wind velocities over 15 mph will blow through most air doors.

c. Display windows.

    1. If window area has no enclosure, measure temperature at first sales space near it. If more than 3° above inside temperature, determine the possibility of increasing return air from this area.

d. Stock areas.

    1. Measure temperature. Determine if these areas can have higher temperature in cooling season.

e. Appliance sales area.

    1. Measure temperature around operating equipment such as televisions. Can heat be removed to return or exhaust for, cooling or recirculated for heating.

f. Vending Machines.

    1. Same as "e".

g. Shipping and receiving department.

    1. Measure air volume going in or out through truck doors. Air curtain at door may not work if negative pressure in building causes opening velocities over 1000 ft/min.

h. Operation periods.

Investigate the possibility of pre-cooling the store before occupancy for perhaps an hour and shutting the system down until there are sufficient people in the store to warrant turning the cooling back on again. You may be able to save approximately 2 hours of operating time a day by this function. Similar possibility is available in the winter time if the store has a considerable number of people and is expected to have this number of people for a period of time, the internal heat from people and lights probably will take care of the store allowing the heating equipment to be shut down.

i. Beauty Parlor.

    1. Ventilation.

    Beauty parlors require considerable ventilation because of the chemical used in treating hair. This particular material is corrosive and any device used for cooling this area should be corrosion resistant such as stainless steel or aluminized steel.

    2. Generated heat.

    Generated heat, either by hair dryers or other devices, can be circulated into the space in the winter time but should be exhausted to the outside in the summer time to reduce the cooling requirement.

j. Alteration department.

The alteration department usually consists of one or trio steam presses, ironers, sewing machines, puffers and similar equipment. High pressure steam is required on the pressing machine and the area must be ventilated because of excessive heat in the summer time. This area should be isolated from the rest of the store by reasonable tight partitions so that air exhausted from this department does not necessarily have to come from the store, although personnel may require some cooling. If the steam boiler is an individual boiler for this department, it is probably oil or gas fired and should be checked for combustion. If there is a restaurant in the store, the steam generation may be supplied from a central boiler used for this purpose. Examine the condensate return system so that condensate is not wasted from the press and dumped in the sewer.

2. Office Buildings.

    a. Entrance Doors – see Department Stores

    b. Elevators – see Department Stores.

(1) If there is more than one elevator, check activity mid-morning, mid-afternoon, evening for possibility of shutting down one or more for those periods. Motor generator should be shut down also.

c. Check method of perimeter area cooling, Is it separate from core system? Is it zoned for direction and solar exposure? Is the air supplied over the outside windows? The latter any cause down drafts over wall cabinets.

d. If there is an atrium, is air supplied at top? Or at various levels? Measure air temperature on walkways on every floor. Cooling may be satisfactory, heating may not. Reduce cooling and lighting during janitorial cleaning. Put on time cycle.

[EXPERIENCE] Cooling and lighting during janitorial cleaning cost $300,000.00 per year in two buildings totalling 560,000 sq.ft.

3. Restaurants

a. Kitchens.

(1) Refrigeration: Inspect condensers for dirt and direction of heated air. Air should be directed away from personnel

Measure outside surface temperature of refrigerator or walk in walls. Cold surface means bad insulation.

(2) Refrigerated display cabinets: Shut off all cooling supply to this area. Air circulation increases cooling loss from open top units.

(3) Check A/C and Refrigeration condensing system for installation of hot gas water heaters.

(4) Check kitchen exhaust hood for maximum of 150 ft per min. cross section air velocity. If more than one exhaust fan and hood, can one be shut off on periods of no cooking? Can a recirculating hood be used?

Exhaust from ranges and other cooking devices should be confined to the cooking surface and be of such size and velocity as to remove the cooking odors and steam. Exhaust hoods may remove sufficient air to be equivalent to as much as 10 times the heating and cooling requirement of that space without exhaust. It is extremely important in the reduction of energy to use "back exhaust" rather than using an overhead hood. Provide a master switch at the entrance door to the room, with a pilot light, so that the last person leaving can shut off all the electric equipment in the room.

(5) Steam kettles should be provided with covers and the steam jacket should be carefully checked for proper traps and adjustment of the relief valves and vents. Any steam venting, or leakage from the lids, should be exhausted directly to the outside through a hood. Shelf type steamers should be provided with front doors and be reasonably tight; also vented into an exhaust hood.

(6) Steam tables add considerably to the cooling requirements of a dining facility. The trays should be reasonably tight fitting and there should be a means of removing the steam and the heat as close as possible to the table rather than allowing it to raise up and distributing to the space, even to a hood overhead. Also check the temperature of the water in the table, or the steam that is used to keep it as close as possible to the boiling point.

(7) Dishwashers

Exhaust hoods at each end should be close to dishwasher opening and large enough to collect all steam. Check for arrangement and exhaust velocity.

Booster water heaters may be self-contained with the washer, or may be provided with higher temperature water from the domestic water system. Examine the operation of this device, as to its efficiency, and to whether the high temperature water flow shuts off after each wash or rinse cycle.

Use of detergents in certain operations will permit lower temperature water and less hot water in certain cycles in a dishwasher, reducing energy use.

Examine the operation of the dishwasher to determine if the washer can be only operated with full loads, rather than allow it to operate with a light load of dishes or utensils. Dishwasher water flow and exhaust should be shut off entirely when dishes are not being cleaned.

(8) Sinks

Since kitchen sinks normally are supplied with 180° water, for sterilizing of pots, pans, make sure these faucets are shut off when not used.

4. Laundries

   a. Washers

      (1) Water flow control

Laundry washers are designed and manufactured with quick opening water valves, so that the washer can be filled with hot water as quickly as possible. These are sized for a 20 second fill period which sometimes greatly overloads the water heating system, unless there are very large storage tanks in use. Sudden removal of high temperature water, which is in the neighborhood of 180°, may affect the operation of the boiler plant by the sudden draw of steam, if the heater is steam heated, or by the excessive operation of fuels, if direct fired. It is suggested that balancing cocks be put on the hot water supply line to the washer to increase this fill time to at least one minute, and preferably 1 1/2 minutes.

Experience has indicated that this small additional amount of time will not upset the overall operation of the laundry.

      (2) Detergent Use

Investigate the possibility of using cold water detergents for those areas which do not require high temperature sterilization. Although detergents are more expensive, it may be cheaper to use this kind of material for cleaning than the high temperature hot water normally used.

      (3) Waste Water Heat Recovery

If the drain arrangement can be properly made, the hot water dumped out of the washer, though dirty and soapy, could be run through a tank in which coils could be installed to preheat the cold water going into the water heater with this waste water. Nearly 30% of the heat necessary to heat the water, perhaps, could be recovered in this manner.

(4) Time Cycle

Determine if the washers are fully loaded for each cycle and whether the washer operation can be set aside until sufficient clothing is available for washing at a full load.

b. Dryers

(1) Temperature Control on Dryers

Temperature control of dryers is essential, not only for energy use but for the condition of the materials being dried, since certain synthetic fabrics are seriously affected by high temperatures. Investigate the possibility of operating at a lower temperature for a longer period of time.

(2) Time-Cycle Dryers

Check to see if the dryer is completely full during each cycle, rather than using it partially filled which will waste fuel for heating.

(3) Temperature Control on Ironers

Ironers, either direct fired or steam heated, should have accurate temperature control, not only to prevent energy waste but to protect fabrics being handled.

(4) Time Cycle on Ironers

Ironers, especially the flatwork type, should be operated only when sufficient material is available for use. Ironers should not run continuously, or be heated when there is no cloth or other materials to be ironed in the machine.

(5) Exhaust and Air Supply

If supply air is not cooled, air should be directed to workers to provide as much air velocity possible for body cooling. Air velocity at person should be at least 300 feet per minute.

5. Supermarkets

    a. Distribution of air to refrigerated display units: Locate the sections of the store which have upright refrigeration cabinets which are open. The air supply system should be directed away from this area. This will result in a fairly cold surrounding area at this location but should not be too uncomfortable to the customers. Determine air flow also across horizontal freezer cabinets, especially with ice cream and similar products, to determine if the air is sweeping these areas and causing additional refrigeration and additional pre-cooling from such areas. This refrigerated equipment is normally in the center of the store and the air outlets could be arranged to deflect the air from these areas.

    b. Service Areas: Refrigerated meat, vegetable, storage and receiving departments. Note that most supermarkets are approximately 12 ft. to 15 ft. high. If these service areas are cooled from the air conditioning system or have separate lower temperature refrigeration systems for air circulation, we suggest that the air supply be lowered to a maximum of 10 ft. and the proper air outlets used so that the upper part of the space does not have to be air conditioned. A layer of warm air near the ceiling can remain which would reduce the cooling requirements for these areas. Check also to determine if there can be some time cycle control on these areas so that they are not cooled during unoccupied periods.

6. Education Facilities:

    a. For administrative areas, offices should be controlled in the same manner as office buildings. For classrooms, each classroom should have a separate temperature control because of the extreme fluctuation in occupancy and heating and cooling requirements. Classrooms should be put on stand-by control, either night shut down or shut off completely, as soon as classes are over and the room is not occupied. It is suggested that if the teacher has an extended period of work after classes are over, that a teacher's study room be provided so that the classroom can be completely shut down without causing the teacher discomfort. Requirements for heating or cooling a classroom for a single person is nearly as much as for an entire class. In auditoriums and assembly halls, due to the rapid change in cooling and heating requirements, oversized equipment is usually necessary. It is suggested that the air be pre-cooled or pre-heated for at least one-half hour before occupancy and be immediately shut down after use is completed. It is also suggested that two-speed equipment be used or other means of capacity reduction which will permit the system to operate at minimum capacity during partial occupancy and at full capacity during complete occupancy. Laboratory and shop facilities should be operated in the same manner as classrooms. Shut off hood exhaust when not used.

b.   Dining facilities: Dining rooms should be controlled separately from kitchens and, if they are isolated into general dining rooms and executive dining rooms, these should be separately controlled also, Heating and cooling systems should be shut down during non-occupied periods. If the dining room is to be used for study hall purposes, then continuous conditioning is required. If it is used for meal time and intermediate recess or break periods, then the system should be two-speed and set on a time clock to automatically decrease the capacity between occupied periods.

c.   Kitchens: (See Restaurants)

d.   Recreation Facilities: Gymnasiums, locker rooms, drying rooms, toilet facilities should all be individually controlled and provided with intermittent operation during non-occupied periods. For gymnasiums which are not air conditioned, economizer control is essential since outside air cooling is required during periods of peak occupancy especially during public sporting exhibitions.

Locker rooms, which require considerable ventilation, can be placed on recirculation or intermittent operation during non-occupied periods. Toilet room exhaust should be interlocked with the supply system to shut off fans during non-occupied periods.

e. Living and Sleeping: Sleeping rooms should by individually controlled but, if not practical, should be at least zoned in accordance with solar exposure or direction. This is also true of wind exposure which has considerable effect, primarily in the wintertime. Instructions should be made to occupants that if they desire to open windows that the heating or cooling system should be shut off during that period so that energy is not wasted due to the effect of outside air. Living rooms should be separately controlled and placed on time clocks to reduce the cooling or heating requirements either by cycling equipment or by changing the temperature during periods of low or non-occupancy. night shut-down or lowering of temperature during winter is important in this application.

f. Laundries: (See 4.)

g. Workshops: Workshops should be controlled in the same manner as classrooms. If exhaust is used in the space, for instance in woodworking, painting and welding areas, these exhaust systems should be interlocked with the make-up air supply so that there is an equivalent amount of air added to that amount exhausted. When the exhaust system is shut down, the outside air supply should return to minimum position required for ventilation.

h. Toilets and Showers: Toilets and showers are required to have a certain amount of exhaust in accordance with building codes and health requirements. A study of the use of the toilet room, especially in dormitories or in large toilet areas, should be made to determine the peak periods of use and the possible reduction in exhaust during non-occupied periods, either by time clocks or by slowing the exhaust fan down. If the toilet is a single person occupancy, the exhaust fan should be interlocked with the light fixture so that the fan only runs when the lights are turned on. If the exhaust system is a central system, a damper can provide the same function as shutting off the fan with a volume control on the exhaust fan, if large size and connected to a number of air outlets.

Temperature in toilet rooms in summer should be warmer than the surrounding building and is normally cooled indirectly by exhaust air from the general surrounding space. In winter time the same is true if the toilet is on an inside area and is heated by means of the exhaust from the surrounding area. In any event, the temperature should be at least 5° cooler in winter and 5° warmer in the summer. This will also reduce occupancy periods by personnel.

7. Hotels:

a. Lobbies, restaurants, laundries – See previous analysis.

b. Guest Rooms: Check thermostat for minimum control temperature. Should have low limit of 72°. Test wall above cabinet units for mildew or moisture. Examine toilet exhaust for time cycle. Exhaust duct should have a backdraft damper.

Test corridor temperature and humidity. Dew point must be lower than guest room temperature.

Empty guest rooms should be cooled at least two hours per day.

J. SPECIAL FACILITIES

a. Hospital Surgeries: Present codes permit 75% recirculation during unoccupied periods. To convert an existing 100% outside air system to partial recirculation, individual surgeries are usually provided with a local fan and dampers mixing supply and return air with a local switch. Check percentage of outside air by comparing supply, return and mixed air temperatures. Find out who changes system and if there is any indication which system is in use.

Examine thermostat for dust. Powder from surgical gloves can affect operation.

b. Hospital Patients Rooms: To get individual room control, these rooms have reheat from central system, fan coils or induction units, Examine air patterns for effect of curtains around beds. If ceiling outlets are used, consider baffles or redirecting air.

c. Computer Rooms: Since this equipment is seriously affected by temperature, air conditioning should be provided by a complete separate system including refrigeration. This will allow other areas to be shut down during unoccupied periods. If there is uneven. temperature, measure air volumes at outlets especially at the furthest location from unit. If under floor supply, look for cable interference.

d. Electronic Clean Rooms: Since these rooms must be under positive pressure, measure exhaust and outside air supply. Check total air volume and supply temperature for humidity control. Examine walls and doors for leaks.

e. Laboratories: Because of contaminants, exhaust hoods are used. Measure face velocity at hood which should be a minimum of 150 ft/min. Consider limiting opening to still use hood and reducing exhaust.

K. SPECIAL EQUIPMENT

a. Thermal Storage: If use of cooling is intermittent, storage of cooling will permit operating smaller equipment for a longer time. If operation can be off peak, utility companies may provide a low rate. Measure compressor operation since suction temperatures are 25° or less. At these temperatures, power input may reach 1.3 kw per ton. Limit the ice thickness on cooler.- Ice is a good insulator and may increase power input if too thick.

b. Heat Wheels: These devices are used with high outside and exhaust air columns. Surface condition may affect humidity removal. Measure static pressure drop and leakage around wheel. Poor filtration may partially plug up openings in wheel. Measure dry and wet bulb temperatures in and out of both air supply and exhaust to determine efficiency. Examine exhaust for possible contamination of air supply.

c. Heat Exchangers: This equipment works in same application as heat wheels except for moisture removal measure temperature in and out of both exhaust and air supply. Check static pressure drop. Compare cooling saving with additional fan power to over come pressure drop.

d. Hot Gas Water Heaters: If domestic water or reheat control can use this waste heat, these heaters have merit. If hot water is used for washing, a special double tube heater is required to prevent contamination if a leak occurs.

e. Management Systems: Operation and control management systems are a good energy conservation application. Examine what the system is doing, whether the operators know what it is supposed to do. Compare read out with actual measurement of space or cooling equipment conditions. Make sure humidity is indicated or controlled. Read out of data may provide enough information for analysis without further field measurement.

f. Well Water Cooling: If well water is below 58° it can provide some or all cooling. To provide dehumidification, six to 10 row tube coils are used. If water is used for condensing, return wells should put warm water in upper part of strata. chemicals in water can reduce capacity by coating tubes or corrosion. If of high content, use a cleanable heat exchanger. If very corrosive, use metals in exchanger such as titanium.

g. Heat Pumps: Since the refrigerant suction pipe handles hot gas for heating, it must be large enough to provide a reasonable pressure drop. Examine if expansion valve, capillary tube and liquid line can handle reverse flow. With outside air at low temperature,, low suction temperature reduces capacity. Average heating capacity is 4,000 BTU/ton at 40° and 1,000 BTU/ton at 0°. Auxiliary heat is almost always required.

h. Heat Pipes: This is a sealed coil, in which an internal refrigerant condenses and evaporates according to the temperature passing through each section. Measure similar to heat wheel and compare savings with power for fan motor.

i. Gas Engine Driven Compressors: The author has used natural gas engine driven refrigeration compressors and generators since 1956. With proper maintenance the engines have lasted 24 years before replacement.

    Waste heat from cooling system can be used for domestic water, space heating and absorption cooling. On engines 1,200 RPM or less 12 PSI steam may be generated, At 1,800 RPM or higher about 200° hot water is available.

Chapter 2

Air Conditioning Systems

Check the static pressure across the cooling coil, which varies from .5" up to 1.25", depending on the thickness of the coil, and the heating coils which should vary from .25" to .75", depending on its thickness. If the pressure drop across these coils is excessive, coils are probably dirty and should be cleaned. Also, on examining the coils, look at the space between the coil and the housing to make sure this is airtight. If air leaks around the coil, this is interfering with the cooling and heating capacity of the coils and the by-passed air is not conditioned. Carefully seal the space around the coils.

Examine also the filters which, if plugged up, will reduce the air quantity. Pressure drop across the filters , should vary from .25" to .45". Most 2" thick panel filters should not exceed .35", bag type filters may go up to .65".

Next, check the entering mixed air temperature.

If excessive outside air is brought in either summer or winter this temperature will increase the requirement for cooling or heating. Determine the minimum outside air which can be brought in to take care of the building. Certain State Code requirements are for so many CFM per person in the building. Code requirements for your particular area should be checked. The CFM times the number of people should give you the ventilation requirement. If there is exhaust in the building for toilets, heat generating equipment and any other devices, the air quantity coming in the outside air may be required to be increased by this amount.

Note that the physical position of an outside air or return air damper does not indicate the air quantity going through the damper. A damper position of 10% physically open may be passing as much as 50% air through this damper. It is essential that actual measurements be made across these openings.

Many systems do not have return air fans or relief openings if there are means of removing air from the building by leakage or if the return air system is short enough that the main air supply fan can draw air from the building. If there is no return fan, there is a tendency to draw more outside air because of the lower restriction in the outside air ductwork than the return air opening or ductwork. More care must be used in reducing the outside air intake volume. If there is excessive outside air coming in, even if the outside air damper is full closed, then a restriction should be placed in the outside air intake by the use of a perforated plate or by closing off part of the duct. Do not reduce the size of the actual outside air intake louver or opening as velocities through this opening, if they exceed 1,000 feet per minute, have a tendency to draw rain and snow into the duct system, which will leak out through the duct joints and damage the inside of the building.

The wet bulb temperature of the cold deck indicates the efficiency of the cooling coil. The closer the wet bulb temperature to the cold deck temperature, the more moisture is removed from the system and the relative humidity in the room air should be reduced. Note that the less moisture removed, the more actual sensible cooling is derived from the same refrigerant system and if the dry air is not necessary, the leaving moisture removal can be reduced. This will allow the refrigerant temperature in the cooling coil to be raised, either chilled water or refrigerant. The higher the refrigerant temperature, the more efficient the refrigeration machine becomes and the less energy will be used.

If it is desired to calculate the total cooling capacity of the system, determine the total heat value, the cold deck wet bulb condition, subtract mixed air total heat value which will give you the total heat of BTU per pound of air which is being removed from the system. Using the standard formula for this purpose, you can then determine the BTU per hour total cooling capacity of the system. This will also determine the amount of refrigerant capacity necessary. There is a slight additional capacity requirement for the refrigerant system since the motor energy going into the supply fan also goes into the air system. The KW input to the fan should be added in the amount of 3,412 BTU per KW per hour to the cooling requirement.

Determining the total heating capacity of the system uses the volume times the hot deck temperature out minus the mixed air temperature using the standard formula for this purpose. This heating capacity will then determine the amount of steam, hot water or electricity necessary to provide the heat. Again, in the heating cycle, examine the outside air quantity and reduce this to the minimum amount required for heating.

Although a humidifier is not shown on the diagram, no field data is indicated for this device. Humidifiers are rarely used in commercial or industrial buildings unless there is a specific process requirement or health requirement. Humidity is usually added by means of a water spray or steam and is controlled by a room humidistat. Since energy is required to generate the steam or evaporate the water in the air system, it is recommended that these systems not be used unless absolutely necessary. If humidity is required, caution should be used as to the percentage of relative humidity as a function of outside temperature. If windows are single pane, at an outside temperature of zero degrees, only 10% relative humidity can be obtained in the building as the excess moisture will condense on the windows. Moisture should not be used to increase the relative humidity in the building if it condenses on any windows or cold wall surfaces.

OTHER FIELD EXAMINATIONS

If possible, examine the joints on the supply ductwork, especially to determine if there is leakage. Dirt streaks at joint corners will show leakage. If the ductwork is poorly assembled, it is possible to lose as much as 60% of the air supply into the space surrounding the ductwork, usually above a ceiling, before it enters the conditioned space outlets. If the ceiling space is a return, this air, of course, returns back to the system. If excessive leakage occurs, this can be partially determined by the temperature of the return air coming back to the air conditioner. If return air is below the average room temperature, short-circuiting is occurring. If there is a lot of recessed lighting in the building and the return air comes through the ceiling space, even with return air temperature equal to the room temperature, there is leakage. This leakage should be reduced by cleaning the joints and taping with duct tape or caulking with an approved material.

Another method of determining air leakage is to measure the air quantity supplied from all the supply air outlets, adding these up, and comparing the total air quantity from the outlets with the air quantity being supplied from the system indicated on the diagram.

Damper motors should close dampers at 90% and 10% of stroke to assure tight fit.

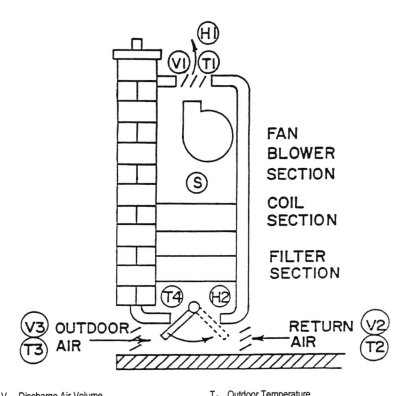

FAN
BLOWER
SECTION

COIL
SECTION

FILTER
SECTION

V_1 Discharge Air Volume
  Air velocity X sq. ft. of discharge grille

T_1 Discharge Temperature
  Winter heating   130°F max.
  Winter cooling   55°F min.
  Summer cooling   55°F min.
  Summer ventilation – room temperature

V_2 Return Air Volume
  Winter heating – Summer cooling
  75% or more of V_1

T_2 Room Temperature

V_3 Outside Air Volume
  Winter heating – Summer cooling
  25% of V_1 or less
  Winter cooling may be up to 100% of V_1.
  Night winter operation – 0% of V_1.

T_3 Outdoor Temperature
  Below 50°F:
    Heating – V_3 at minimum
    Cooling – V_3 variable to provide T_1 at 55° – 60°F.
  50°F to 60°F:
    Heating – V_3 at minimum
    Cooling – V_3 variable to provide T_1 at 55° – 60°F.
  Above 60°F:
    Heating-Cooling – V_3 at minimum.

T_4 Mixed Air Temperature

S Static Pressure
  If more than .25 inches of water, examine coils
  and filters for dirt.

H_1 Supply Air, H_2 Mixed Air wet bulb temperature

V_1 _____
V_2 _____
V_3 _____
S _____

T_1 _____
T_2 _____
T_3 _____

T_4 _____
H_1 _____
H_2 _____

Figure 2-1 AC 1 / Unit Ventilator / Unit Air Conditioner
Field Measurements

AC1
UNIT VENTILATOR
UNIT AIR CONDITIONER

Since the room thermostat controls (TI), and the system supply temperature to the room is in proportion to the room requirements, this temperature will vary with the room thermostat position. In order to determine the maximum cooling capacity, if unit is an air conditioner, the thermostat should be set at least 5 degrees below the room temperature which should place the cooling system in maximum capacity.

In the winter season, in order to find the maximum heating capacity, thermostat should be set at least 5 degrees above room temperature.

If the supply air volume (V1) is too low, in order to supply sufficient cooling, supply temperature may be lower than 55 degrees in summer and may be higher than 130 degrees in the winter. If these temperatures are experienced, check to find the actual original capacity of the system. In any event, if these temperatures are required to satisfactorily take care of the space, the air quantity should be increased.

Fan suction static pressure measurements are used to find out whether there are any restrictions in the coils or filters which is causing this reduction in air quantity. The negative static pressure at the fan inlet should not exceed .5 inches. Remove the front cover and inspect the coils and fan for dirt.

Examine the filters which, if plugged up, will reduce the air quantity. Pressure drop across the filters should vary from .10" to .20".

Check outside air volume again when there is wind velocity against the outdoor intake. Provide a restricting perforated plate in intake if outdoor air quantity is too great.

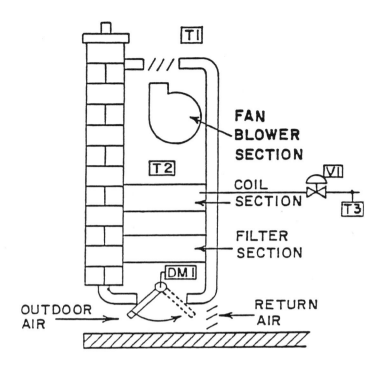

T1 <u>Room thermostat</u>

Winter – room temperature below T1 setting, T1 opens hot water or steam valve V1 or energizes electric heater. Outdoor damper DM1 set at minimum position ± 20% total air supply.

Room temperature above T1 setting, V1 closes, outdoor air damper DM1 gradually opens, interlocked return air closes.

T2 <u>Low limit thermostat</u>

On winter cooling T2 set 55° to 60° overrides T1 to prevent supply air temperature below T2 setting.

Summer cooling – if coil is used with chilled water, action of T1 is reversed by water line aquastat T3 so T1 opens V1 for cooling with outdoor air damper DM1 at minimum position.

Figure 2-2 AC1 / Unit Ventilator / Unit Air Conditioner
Typical Control

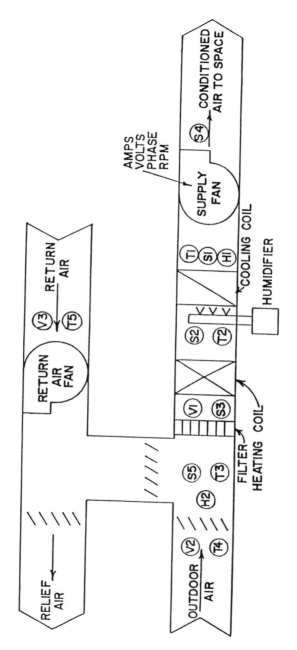

Figure 2-3   AC 2 / Single Zone Air Conditioner
Field Measurements

AC2
SINGLE ZONE AIR CONDITIONER
FIELD MEASUREMENTS

T1	System Supply Temperature	T1 _____
	Summer, variable from room stat 52° minimum.	T2 _____
	Winter, variable from room stat 130° maximum.	
		T3 _____
T2	Heating Air Temperature	T4 _____
	Summer T2 = T3	
	Winter T2 = T1	T5 _____
T3	Mixed Air Temperature	H1 _____
	T3 = % T5 + % T4	
	Night – T3 = T5	H2 _____
T4	Outdoor Air Temperature	V1 _____
T5	Return Air Temperature	V2 _____
	Room temperature ± 3°	V3 _____
H1	Supply Wet Bulb (humidity) Temperature	S1 _____
	Summer H1 = T1 = ± 3°	
	Winter – variable	S2 _____
H2	Mixed Air Wet Bulb Humidity & Temperature	S3 _____
V1	Total Air Supply Volume	S4 _____
	Measure leaving side of filters.	S5 _____
V2	Outside Air Volume	AMPS _____
V3	Return Air Volume	Volts _____
S1	Suction Side Static Pressure	Phase _____
S4	Discharge Side Static Pressure	RPM _____
S1 + S4	System Static Pressure	
S1 – S2	Cooling Coil Pressure Drop	
S2 – S3	Heating Coil Pressure Drop	
S3 – S5	Filter Pressure Drop	

AC2
SINGLE ZONE AIR CONDITIONER

Since the room thermostat controls (T1), and the system supply temperature to the room is in proportion to the room requirements, this temperature will vary with the room thermostat position. In order to determine the maximum cooling capacity, the thermostat should be set at least 5 degrees below the room temperature which should place the cooling system in maximum capacity.

In the winter season, in order to find the maximum heating capacity, thermostat should be set at least 5 degrees above room temperature.

If the system volume (V1) is too low, in order to supply sufficient capacity, supply temperature may be lower than 55 degrees in summer and may be higher than 110 degrees in the winter. If these temperatures are experienced, check to find the actual original capacity of the system In any event, if these temperatures are required to satisfactorily take care of the space, the air quantity should be increased.

Field static pressure measurements are used to find out whether there are any restrictions in the air system which are causing this reduction in air quantity. The static pressure across the supply fan is (S1 + S4) (add the two figures together even though one is negative pressure and one is positive pressure).

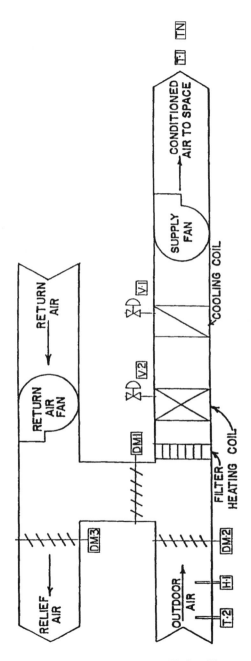

Figure 2-4 AC 2 / Single Zone Air Conditioner
Typical Control

AC2
SINGLE ZONE SYSTEM
TYPICAL CONTROL

Manual supply fan start, return fan, if used, runs also.

Room Stat [T1]   2° below setting [T1] operates valve [V2] on heating coil, steam, hot water or electric relay. The lower the room temperature the more heat supplied.

[T1] may be two stats – one heating, one cooling.   2° above setting [T1] operates chilled water or refrigerant solenoid valves [V1] on cooling coil. The higher the room temperature, the colder the supply temperature, minimum usually 52°F.

[DM1] Return air and [DM2, 3] Outdoor air and relief dampers operate at 90° from each other.
Night cycle heating only. [TN] operates [V2] and starts and stops supply fan. [DM2] and [DM3] closed. Time cycle 2 starts maximum per hour.

Economizer: [T2] Outdoor air stat above 60° on heating and below 50° on cooling locks outside air and relief dampers in minimum position. Below 60° outdoors [T2] locks out [V1] and thermostat [T1] modulates outdoor and relief dampers open for cooling. If outdoor humidity is too high, [H1] locks out [T2], [T1] again operates [V1].

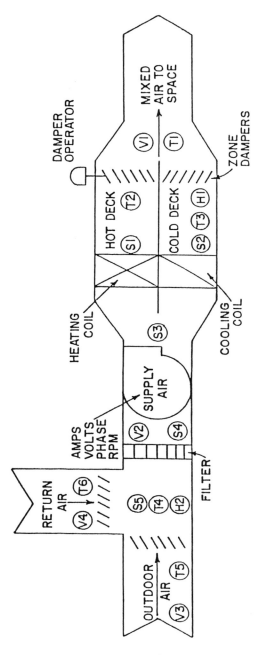

Figure 2-5 AC 3
Multizone Air Conditioner / Hot and Cold Deck
Field Measurements

AC3
MULTIZONE AIR CONDITIONER
HOT AND COLD DECK
FIELD MEASUREMENTS

T1 Zone Supply Temperature
Full heating T1 = T2
Full cooling T1 = T3

T2 Hot Deck Temperature
Heating – variable 130° maximum
Cooling T2 = T4

T3 Cold Deck Temperature
Heating T3 = T4
Cooling minimum 52°

T4 Mixed Air Temperature
T4 = % T5 + % T6

T5 Outdoor Air Temperature

T6 Return Air Temperature
T6 = average room temperature ± 3°

H1 Wet Bulb Temperature (humidity)
T3 – H1 = ± 3°

H2 Mixed Air Wet Bulb Humidity & Temperature

S1 – S3   Pressure Drop – heating coil

S2 – S3   Pressure Drop – cooling coil

S3 – S4   System Static Pressure

S4 – S5   Pressure Drop, filters

V1 Zone Air Supply Volume

V2 System Air Volume
Measure leaving side of filters

V3 Outdoor Air Volume

V4 Return Air Volume

T1 _____
T2 _____
T3 _____
T4 _____
T5 _____
T6 _____
H1 _____
H2 _____
S1 _____
S2 _____
S3 _____
S4 _____
S5 _____
V1 _____
V2 _____
V3 _____
V4 _____
AMPS _____
Volts _____
Phase _____
RPM _____

AC3
MULT I Z ONE AIR CONDITIONER
HOT AND COLD DECK

Since the room thermostat controls (T1), and the zone supply temperature to the room is in proportion to the room requirements, this temperature will vary with the room thermostat position. In order to determine the maximum cooling capacity, the thermostat should be set at least 5 degrees below the room temperature which should place the cold deck dampers in the full open position. If (T1) is more than 2 degrees higher than (T3), hot deck damper leaks – adjust mixing damper.

In the winter season, in order to find the maximum heating capacity, thermostat should be set at least 5 degrees above room temperature. This will place hot deck damper full open, If (T1) is more than 5 degrees below (T2), cold deck damper leaks – adjust mixing damper.

Tightness of mixing dampers is most important. Check for loose damper linkage.

If the zone volume (V1) is too low, in order to supply sufficient capacity, cold deck temperature (T3) may be lower than 55 degrees in summer, and hot deck temperature (T2) may be higher than 110 degrees in the winter. If these temperatures are experienced, check to find the actual original capacity of the system. In any event, if these temperatures are required to satisfactorily take care of the space, the air quantity should be increased. Hot deck temperature (T2) should be varied in accordance with outside temperature, either manually or automatically.

Field static pressure measurements are used to find out whether there are any restrictions in the air system which are causing this reduction in air quantity. The static pressure across the supply fan is (S3 + S4) (add the two figures together even though one is negative pressure and one is positive pressure).

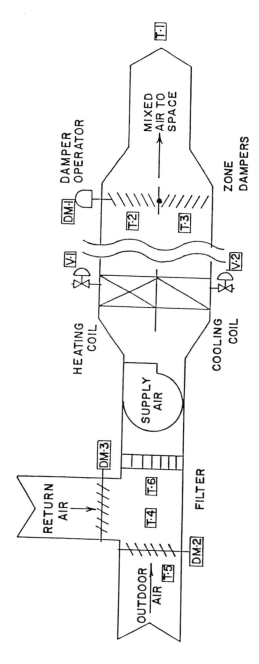

Figure 2-6   AC 3 / Multizone Air Conditioner / Hot and Cold Deck
Typical Control

AC3
MULTIZONE AIR CONDITIONER
HOT AND COLD DECK
TYPICAL CONTROLS

T1   Room Thermostat
      T1 operates mixing damper DM1 in each zone to provide supply
      air temperature to match thermostat setting.

T2   Hot Deck Thermostat
      Summer T2 = T4.
      Winter – variable reset from outdoor thermostat T5 operates steam
      or hot water valve V1 or electric contactors.

T3   Cold Deck Thermostat
      Winter T3 = T4
      Summer T3 operates chilled water or refrigerant valves V2

T5   Outdoor Air Thermostat Controls
      Above 55° outdoors, outdoor air damper remains in minimum
      position.
      At outdoor temperatures below 55°F, T4 gradually opens outdoor
      air damper DM2 and closes return air damper DM3 for cooling.
      Where outdoor air temperatures may go below 32°F, heating coil
      shall be located on entering side of cooling coil. T6 freezestat
      shall close outdoor air damper then stop fan to prevent subfreezing
      air leaving heating coil.

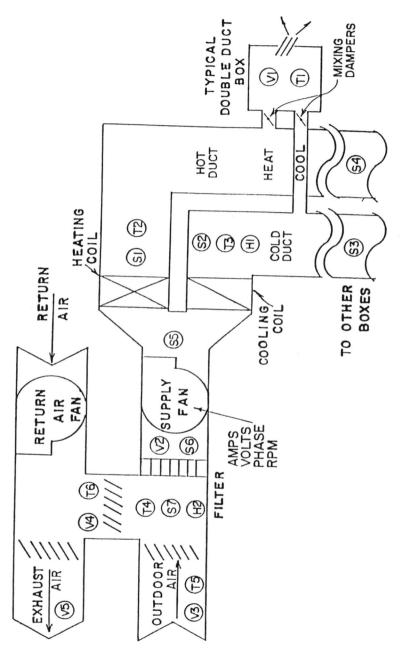

Figure 2-7   AC4 / Multizone Air Conditioning / Double Duct System
Field Measurements

AC4
MULTIZONE AIR CONDITIONER
DOUBLE DUCT SYSTEM
FIELD MEASUREMENTS

T1   Mixing Box or Duct Temperature
Full heating T1 = T2 – 2°
Full cooling T1 = T3 + 2°

T2   Hot Deck Temperature
Winter – 130° max. variable with outside temperature.
summer T2 = T4

T3   Cold Deck Temperature
Summer – ± 52°
Winter T3 = T4

T4   Mixed Air Temperature
T4 = % T5 + % T6

T5   Outdoor Air Temperature

T6   Return Air Temperature

H1   System Wet Bulb Temperature
(Humidity) H1 = T3 ± 3° on full cooling.

H2   System Air Wet Bulb Humidity

V1   Zone Air Supply Volume
Check for variation on full heating vs. full cooling

V2 = V1 + etc. all outlets

V2 = % V3 + % V4

V3   Outdoor Air Volume

V4   Return Air Volume

V5   Exhaust Volume
V5 = V3 – building exhaust

S1   Hot Deck Static Pressure

S2   Cold Deck Static Pressure

S3   Cold Duct Static Pressure at Duct End
S3 = S2 – 1" to 2"

S4   Hot Duct Static Pressure at Duct End
S4 = S1 – 1" to 2"

S5   System Supply Pressure

S6   System Suction Pressure

S5 + S6 = System Total Pressure

S6 – S7 = Pressure Drop – filters

S5 – S1 = Pressure Drop – heating coil

S5 – S2 = Pressure Drop – cooling coil

T1 _____	V1 _____	S4 _____
T2 _____	V2 _____	S5 _____
T3 _____	V3 _____	S6 _____
T4 _____	V4 _____	S7 _____
T5 _____	V5 _____	AMPS_____
T6 _____	S1 _____	Volts_____
H1 _____	S2 _____	Phase_____
H2 _____	S3 _____	RPM_____

AC4
MULTIZONE AIR CONDITIONER
DOUBLE DUCT SYSTEM

Since the room thermostat controls (T1) by positioning dampers from hot and cold ducts so that the temperature to the room is in proportion to the room requirements, this temperature will vary with the room thermostat position. In order to determine the maximum cooling capacity, the thermostat should be set at least 5 degrees below the room temperature which should place the cooling duct damper full open and heating duct damper closed. Outlet air supply should be not more than 2 degrees above (T3). If higher, hot damper may leak.

In the winter season, in order to find the maximum heating capacity, thermostats should be set at least 5 degrees above room temperature. If air supply is more than 2 degrees below (T2), cold damper may leak.

If supply air volume at outlet is not sufficient to heat or cool, increase volume with manual take-off dampers. If this doesn't work, increase static pressures (S3) and (S4). (See note on Mixing Boxes.)

If the system volume (V2) is too low, in order to supply sufficient capacity, cold supply temperature (T3) may be lower than 55 degrees in summer and hot supply temperature (T2) may be higher than 110 degrees in the winter. If these temperatures are experienced, check to find the actual original capacity of the system. In any event, if these temperatures are required to satisfactorily take care of the space, the air quantity should be increased.

Field static pressure measurements are used to find out whether there are any restrictions in the air system which are causing this reduction in air quantity. The static pressure across the supply fan is (S5) + (S6) (add the two figures together even though one is negative pressure and one is positive pressure).

If cold duct and hot duct are long and there are numerous mixing dampers or box take-offs, there may be considerable variation in static pressures. This can be determined by measurements (S3) and (S4) near the ends of these ducts. Variation in pressure will cause changes in room outlet volume interfering with the controls.

Correction of this condition requires dampers in hot duct at (S1) and cold duct at (S2) operated from duct static pressure controllers at (S4) and (S3), respectively.

If pressure in ducts varies too much, cold air will leak into hot duct in summer and hot air into cold duct in winter.

If the system is low pressure, zone air supply to outlets is provided by dampers in duct connections to each of the hot and cold ducts by working opposite each other. Dampers should be tight fitting at full open or full closed. Set damper motor operation similar to outside and return air dampers.

If the system is medium pressure (up to 10 inches water), these dampers may be in a mixing box which not only controls supply temperature but reduces the main supply pressure to a lower, quieter supply pressure. Adjustment of these dampers is also important. These dampers work differently than the mixing dampers. Check the manufacturer's instruction for adjustment.

Volume adjustment on a mixing box may be by an automatic static pressure damper. Adjustment for increasing volume should be in accordance with manufacturer's instructions.

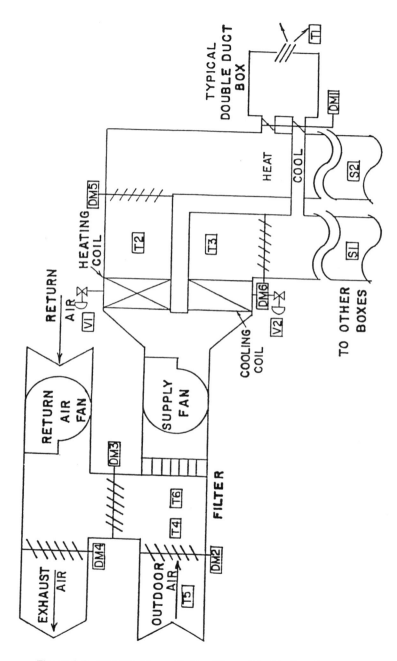

Figure 2-8   AC4 / Multizone Air Conditioner / Double Duct System
Typical Control

AC4
MULTIZONE AIR CONDITIONER
DOUBLE DUCT SYSTEM
TYPICAL CONTROL

T1   Room Thermostat
      T1 operates mixing damper DM1 in each zone to provide supply
      air temperature to match thermostat setting.

T2   Hot Deck Thermostat
      Summer T2 = T4.
      Winter – variable reset from outdoor thermostat T5 operates steam
      or hot water valve V1 or electric contactors.

T3   Cold Deck Thermostat
      Winter T3 = T4
      Summer T3 operates chilled water or refrigerant valves V2

T5   Outdoor Air Thermostat Controls
      Above 55°F outdoors, outdoor air damper DM2 remains in mini-
      mum position.
      At outdoor temperatures below 55°F, T4 gradually opens outdoor
      air damper DM2 and closes return air damper DM3 for cooling.
      Where outdoor air temperature may go below 32°F, heating coil
      shall be located on entering side of cooling coil. T6 freezestat
      shall close outdoor air damper then stop fan to prevent subfreezing
      air leaving heating coil.
      Exhaust damper DM4 interlocked with DM2.

S1   Cold Duct Static Pressure Controller at end of duct adjusts damper
      DM6.

S2   Hot Duct Static Pressure Controller at end of duct adjusts damper
      DM5.

S1 = S2

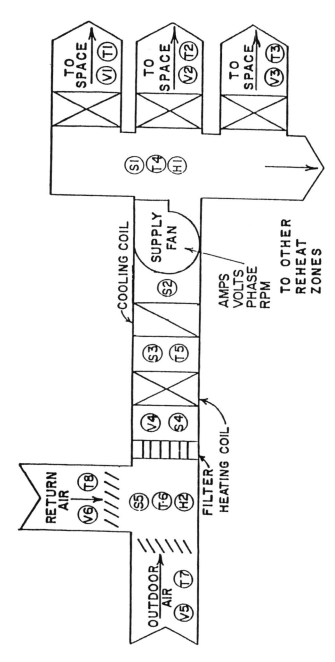

Figure 2-9   AC5 / Multizone Air Conditioner / Zone Reheat
Field Measurements

AC5
MULTIZONE AIR CONDITIONER
ZONE REHEAT

FIELD MEASUREMENTS

T1, T2, T3	Zone Supply Temperature	T1 _____
	Variable from room thermostat.	T2 _____
	Full cooling T1, T2, T3 = T4	T3 _____
T4	System Supply Temperature	T4 _____
T5	Winter T5 = T4	T5 _____
	Summer T5 = T6	T6 _____
T6	Mixed Air Temperature	T7 _____
	T6 = % T7 + % T8	T8 _____
T7	Outdoor Air Temperature	H1 _____
T8	Return Air Temperature	H2 _____
	T8 = Average room temperature ± 3°	V1 _____
H1	Wet Bulb Temperature (humidity)	V2 _____
	T4 − H1 = ± 3°	V3 _____
H2	Mixed Air Wet Bulb Humidity and Temperature	V4 _____
V1, V2, V3	Zone Air Volume	V5 _____
V4 = V1 + V2 + V3 etc.		V6 _____
	Measure leaving side of filters.	S1 _____
V5	Volume Outdoor Air	S2 _____
V6	Volume Return Air	S3 _____
S1 + S2	System Static Pressure	S4 _____
S2 − S3	Cooling Coil Pressure Drop	S5 _____
S4 − S3	Heating Coil Pressure Drop	AMPS _____
S4 − S5	Filter Pressure Drop	Volts _____
		Phase _____
		RPM _____

AC5
MULTIZONE AIR CONDITIONER
ZONE REHEAT

Since the room thermostat controls (T1, T2, T3) and the zone supply temperature to the room is in proportion to the room requirements, this temperature will vary with the room thermostat position. In order to determine the maximum cooling capacity, the thermostat should be set at least 5 degrees below the room temperature which should shut off reheat coil and supply cooling system air at temperature (T4).

Measure air temperature entering and leaving the reheat coil under full cooling to make sure reheat coil completely shuts off. Slight leakage of heating fluid can load up the cooling system.

In the winter season, in order to find the maximum heating capacity, thermostats should be set at least 5 degrees above room temperature, which will open reheat coil to full temperature.

If the zone volume (V1, V2, V3) is too low, in order to supply sufficient capacity, main supply temperature (T4) may be lower than 55 degrees in summer and may be higher than 75 degrees in the winter. If these temperatures are experienced, check to find the actual original capacity of the system. In any event, if these temperatures are required to satisfactorily take care of the space, the air quantity should be increased.

Field static pressure measurements are used to find out whether there are any restrictions in the air system which are causing this reduction in air quantity. The static pressure across the supply fan is (S1) + (S2). (Add the two figures together even though one is negative pressure and one is positive pressure.

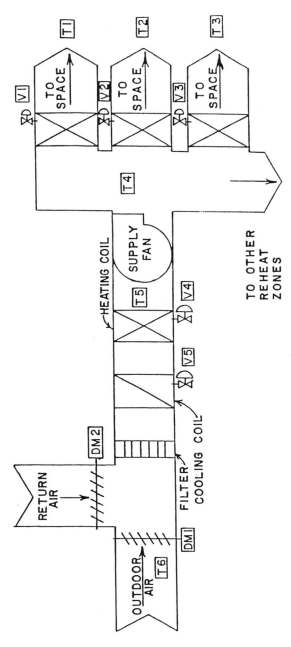

Figure 2-10 AC5 / Multizone Air Conditioner / Zone Reheat
Typical Control

AC5
MULTIZONE AIR CONDITIONER
ZONE REHEAT
TYPICAL CONTROLS

T1, T2, T3   Zone Room Thermostats

Operate steam or hot water valves V1, V2, V3 or electric contactors.

T4   System Supply Thermostat

Usually constant 52° – 55°F.

T4 opens steam or hot water V4 or electric contactor for heating.

T4 opens chilled water or refrigerant valves V5 for cooling.

T5   Freezestat

Shall close outdoor air damper DM1 and then stop fan to prevent subfreezing air leaving heating coil.

T6   Outdoor Air Thermostat Controls

Above 55°F outdoors, outdoor air damper DM1 remains in minimum position.

At outdoor temperatures below 55°F, T4 gradually opens outdoor air damper DM1 and closes return air damper DM2 for cooling.

Where outdoor air temperature may go below 32°F, heating coil shall be located on entering side of cooling coil.

DM1 Outdoor Air Damper

Operates at 90° to return air damper DM2.

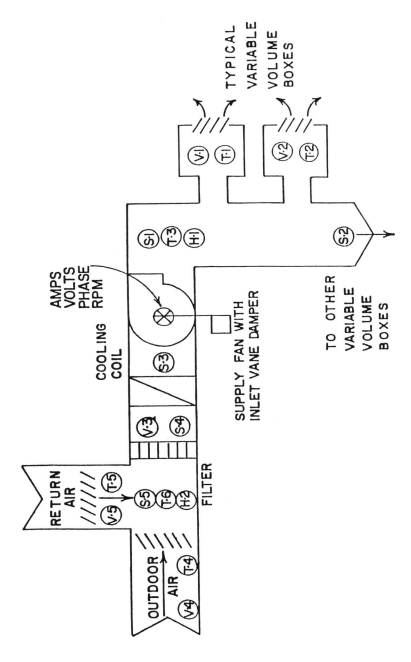

Figure 2-11   AC6 / Variable Volume System
Field Measurements

AC6
VARIABLE VOLUME SYSTEM
FIELD MEASUREMENTS

T1, T2	Supply Temperature Summer – ± 52° Winter – variable to room temperature	T1 _____
		T2 _____
		T3 _____
T3	System Temperature Full cooling T3 = T1, T2 Winter variable to room temperature	T4 _____
		T5 _____
T4	Outdoor Air Temperature	T6 _____
T5	Return Air Temperature Room temperature ± 3°	H1 _____
		H2 _____
T6	Mixed Air Temperature T6 = % T4 + % T5	V1 _____
		V2 _____
H1	Supply Wet Bulb Temperature (humidity) Summer H1 = T3 ± 3° Winter – variable	V3 _____
		V4 _____
		S1 _____
H2	Mixed Air – Wet Bulb Humidity and Temperature	S2 _____
V1, V2	Box Supply Volume Measure at full cooling.	S3 _____
		S4 _____
V3	System Supply Volume V3 = V1 + V2 + etc. at full cooling. Measure leaving side of filters.	S5 _____
		AMPS _____
V4	Outdoor Air Volume	Volts _____
V5	Return Air Volume	Phase _____
S1	System Static Pressure	
S2	Minimum Static Pressure to make boxes work.	
S1 + S3	System Static Pressure	
S3 – S4	Cooling Coil Pressure Drop	
S4 – S5	Filter Pressure Drop	

AC6
VARIABLE VOLUME AIR CONDITIONING SYSTEM

NOTE: The advantage of this system is based on reduced air flow and fan horsepower on lower cooling requirements. Fan volume control device must be functioning properly to obtain this operation saving. (VAV) = Variable Air Volume control box.

Since the room thermostat controls (V1), (V2), and the system supply temperature (T3) is constant, this volume will vary with the room thermostat position. In order to determine the maximum cooling capacity, the thermostat should be set at least 5 degrees below the room temperature which should place the supply volume at maximum capacity.

In order to find the minimum volume, thermostat should be set at least 5 degrees above room temperature. If minimum ventilation is required, provide a stop on the VAV box control.

If the system volume (V3) is too low, in order to supply sufficient capacity, supply temperature may be lower than 55 degrees. If this temperature is experienced, check to find the actual original capacity of the system. In any event, if these temperatures are required to satisfactorily take care of the space, the air quantity should be increased. Check the system static pressure controller whether operating a damper or fan inlet vortex control for full open position. Do not set fan inlet vortex damper at less than 15 degrees to perpendicular to inlet.

Field static pressure measurements are used to find out whether there are any restrictions in the air system which are causing this reduction in air quantity. The static pressure across the supply fan is (S1) + (S3). (Add the two figures together even though one is negative pressure and one is positive pressure.)

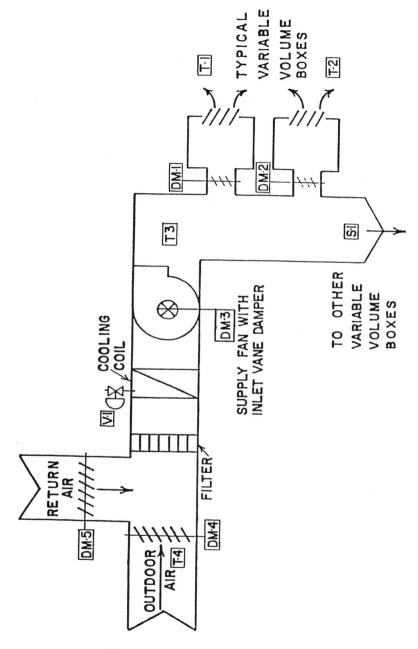

Figure 2-12   AC6 / Variable Volume System
Typical Control

AC6
VARIABLE VOLUME SYSTEM
TYPICAL CONTROLS

T1, T2   Room thermostats adjust DM1, DM2 dampers on air supply to variable volume boxes.

T3   Supply air temperature opens chilled water or refrigerant valve V1 for cooling.

T4   Outdoor air thermostat – Below 55° outside, T3 through T4 operates outdoor air damper DM4 and oppositely return air damper DM5 to keep T3 temperature setting. T3 may also be outside compensated by T4.

S1   Static pressure controller at end of duct gradually closes DM3 vortex damper on fan inlet to reduce air flow.

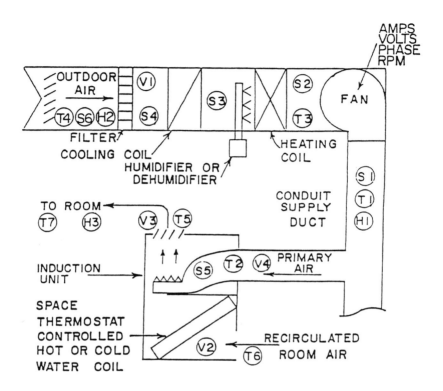

Figure 2-13   AC7 / Induction System
Field Measurements

AC7
INDUCTION SYSTEM

T1 System Supply Temperature
Summer 54° if induction units have no drains.
50° – 52° with drains. Winter variable.

T2 Unit Supply Temperature
T2 – T1 + 2° maximum

T3 System Suction Temperature
T3 = T1 – 2° maximum

T4 Outdoor Air Temperature

T5 Unit Supply Temperature
T5 = % T2 + % T6

T6 Recirculated Room Air Temperature

T7 Room Temperature

V1 System Supply Volume
Measure leaving side of filters

V2 Unit Induced Air

V3 Unit Supply Air

V4 Unit Primary Air
V4 = V3 – V2

S1 System Supply Pressure
3" to 10" water gauge

S2 System Suction Pressure

S1 + S2 System Total Pressure

S2 – S3 Heating Coil Pressure Drop

S3 – S4 Cooling Coil Pressure Drop

S5 Primary Air Static Pressure
Must be enough to operate induction unit.
Measure at nozzles or in supply duct.

S6 Outdoor Air Intake Static Pressure

H1 System Wet Bulb Temperature

H2 Outdoor Wet Bulb Temperature

H3 Room Wet Bulb Temperature

T1 _____
T2 _____
T3 _____
T4 _____
T5 _____
T6 _____
T7 _____
V1 _____
V2 _____
V3 _____
V4 _____
S1 _____
S2 _____
S3 _____
S4 _____
S5 _____
S6 _____
H1 _____
H2 _____
H3 _____
AMPS _____
Volts _____
Phase _____
RPM _____

AC7
AIR CONDITIONING INDUCTION SYSTEM

Since the room thermostat controls (T5), and the water supply volume to the unit coil is in proportion to the room requirements, this temperature will vary with the room thermostat position. In order to determine the maximum cooling capacity, the thermostat should be set at least 5 degrees below the room temperature which should place the chilled water supply valve fully open,

In the winter season, in order to find the maximum heating capacity, thermostat should be set at least 5 degrees above room temperature to open fully the hot water supply valve to coil. (NOTE that the water temperature to the coil has reversed the thermostat operation.)

If the unit volume (V3) is too low, in order to supply sufficient capacity, supply temperature (T5) may be lower than 55 degrees in summer and may be higher than 110 degrees in winter. If these temperatures are experienced, check the primary air volume to the unit (V4) and temperature (T2). System air supply (T1) is normally 52 degrees to 55 degrees in summer and 55 degrees to 60 degrees in winter. Static pressure of primary air (S5) to unit determines the total air induced (V3). There is a minimum pressure permissible for each unit. Check manufacturer's data.

Field static pressure measurements are used to find out whether there are any restrictions in the primary air system which are causing a reduction in air quantity. The static pressure across the supply fan is (S1) + (S2). (Add the two figures together even though one is negative pressure and one is positive pressure.)

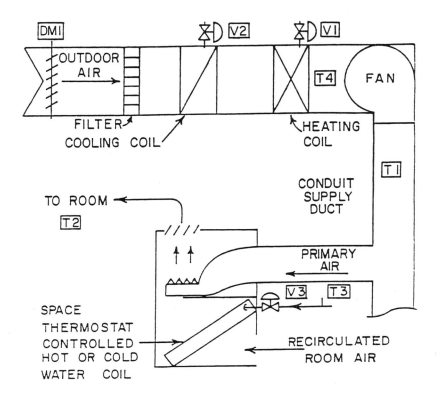

T1   Primary air thermostat set at 55° to 60° in winter, 52° to 55° in summer.

Opens valve V1 on heating coil supply of steam, hot water or energizes electric for increase in temperature.

For decrease in temperature, T1 opens valve V2 on cooling coil supply of chilled water or refrigerant.

Where outdoor air may go below 32°F, heating coil shall be on entering side of cooling coil.

T2   Room Thermostat. Winter T2 opens hot water valve V3 for heating. Summer T2 opens same chilled water valve V3 for cooling.

T3   Water line aquastat reverses operation of T2 by sensing water supply temperature supply – hot or chilled water.

T4   Freezestat will stop fan and close outside damper DM1 if air leaving heating coil goes below 40°F.

Figure 2-14 AC7 / Induction System
Typical Control

Chapter 3

Refrigeration

R1

DIRECT EXPANSION REFRIGERATION AIR COOLED EVAPORATOR AIR COIL

FIELD MEASUREMENTS

SP1	Refrigerant Suction Pressure at Compressor	SP1	_____
ST1	Refrigerant Suction Temperature at Compressor	SP2	_____
		DP	_____
SP2	Refrigerant Suction Pressure at Evaporator	LP	_____
ST2	Refrigerant Suction Temperature at Evaporator	CTI	_____
DP	Refrigerant Discharge Pressure	CTO	_____
DT	Refrigerant Discharge Temperature	CAV	_____
LP	Refrigerant Liquid Pressure	ST1	_____
LT	Refrigerant Liquid Temperature	ST2	_____
CTI	Condenser Air Temperature In	DT	_____
CTO	Condenser Air Temperature Out	LT	_____
CAV	Condenser Air Volume	AMPS	_____
AMPS	Compressor Motor	Volts	_____
Volts	Compressor Motor	Phase	_____
Phase	Compressor Motor		

Type of Refrigerant _____

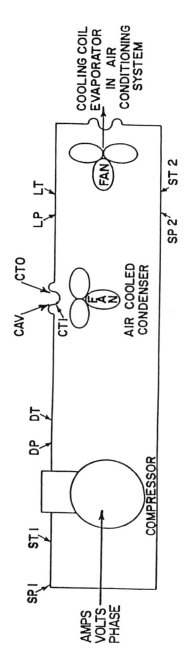

Figure 3-1   R1 / Direct Expansion Refrigeration Air Cooled / Evaporator Air Coil
Field Measurements

R1
REFRIGERATION
DIRECT EXPANSION AIR COOLED EVAPORATOR AIR COIL

A. Compressor suction and discharge pressures are a function of the operation and determine the power input to the compressor motor. The lower the discharge pressure (DP) and the higher the suction pressure (SP1), the less KW the motor requires.

B. Suction temperature (ST1) should be higher than the equivalent refrigerant pressure-temperature so that liquid refrigerant does not return to compressor and cause damage.

C. Discharge gas temperature (DT) is also higher than the equivalent refrigerant pressure-temperature because motor heat is added. If (DT) is not above equivalent temperature, refrigerant system needs examination for refrigerant volume, oil, air.

D. The air cooled condenser changes the refrigerant hot gas to liquid by removing the latent heat of vaporization. Liquid pressure (LP) should be very close to discharge gas pressure (DP). Liquid temperature (LT) should be lower than discharge temperature (DT).

E. (a) Capacity of air cooled condenser is determined by the fan air volume (CAV), the entering air temperature (CTI) and leaving air temperature (CTO).

   (b) If air quantity is low, the coil is dirty, or there is excessive oil in the condenser coil the capacity may be low. Compare BTU/hr. condenser capacity with system cooling capacity plus compressor motor input. See standard formulas.

F. Refrigerant readings at evaporator coil should be made if piping between evaporator coil and compressor is long or if the compressor is several feet above the coil. (SP1 – SP2) indicates refrigerant suction temperature rise which should not be more than 3 degrees. Check pipe insulation. (ST2) should be about 5 degrees warmer than the equivalent temperature of (SP2). If colder, adjust expansion valves. (See System Analysis of Cooling Coil.)

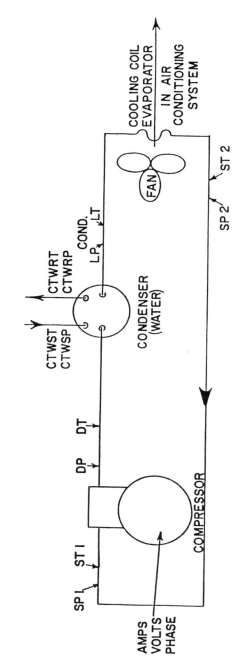

Figure 3-2 R2 / Direct Expansion Refrigeration Water Cooled Evaporator Air Coil
Field Measurements

R2
DIRECT EXPANSION REFRIGERATION WATER COOLED EVAPORATOR AIR COIL
FIELD MEASUREMENTS

SP1   Refrigerant Suction Pressure at Compressor

ST1   Refrigerant Suction Temperature at Compressor

SP2   Refrigerant Suction Pressure at Evaporator

ST2   Refrigerant Suction Temperature at Evaporator

DP   Refrigerant Discharge Pressure

DT   Refrigerant Discharge Temperature

LP   Refrigerant Liquid Pressure

LT   Refrigerant Liquid Temperature

CTWST   Cooling Tower Water Supply Temperature

CTWSP   Cooling Tower Water Supply Pressure

CTWRT   Cooling Tower Water Return Temperature

CTWRP   Cooling Tower Water Return Pressure

AMPS   Compressor Motor

Volts   Compressor Motor

Phase   Compressor Motor

SP1 _____
SP2 _____
ST1 _____
SP2 _____
DP _____
DT _____
LP _____
LT _____
CTWSP_____
CTWST_____
CTWRP_____
CTWRT_____
AMPS_____
Volts _____
Phase _____
Type of Refrigerant _____

R2
REFRIGERATION
DIRECT EXPANSION WATER COOLED EVAPORATOR AIR COIL

A. Compressor suction and discharge pressures are a function of the operation and determine the power input to the compressor motor. The lower the discharge pressure (DP) and the higher the suction pressure (SP1), the less KW the motor requires.

B. Suction temperature (ST1) should be higher than the equivalent refrigerant pressure-temperature so that liquid refrigerant does not return to compressor and cause damage.

C. Discharge gas temperature (DT) is also higher than the equivalent refrigerant pressure-temperature because motor heat is added. If (DT) is not above equivalent temperature, refrigerant system needs examination for refrigerant volume, oil, air.

D. (a) The water cooled condenser changes the refrigerant discharge gas to liquid by removing the latent heat of vaporization. Refrigerant liquid pressure (LP) should be slightly less than discharge pressure (DP). If this pressure drop (DP – LP) is over 5 PSI, check condition of condenser (refrigerant is usually in the shell).

    (b) Condenser cooling water either from cooling tower, wells or city water, usually has a temperature rise (CTWST – CTWRT) of 10 degrees F at full capacity for cooling towers, 95 degrees in, 85 degrees out maximum; for well or city water, 20 degrees F rise if water supply is restrictive or expensive. The lower the return water temperature, the lower the refrigerant discharge pressure, the lower the compressor power input.

    (c) Condenser water pressure drop usually varies around 15 PSI. Higher pressure drop (CTWSP – CTWRP) may indicate restrictions in water tubes.

    (d) Low condenser water temperature rise (CTWST – CTWRT) plus low refrigerant temperature drop (DT – LT) may indicate low refrigerant volume, air in system, excessive oil.

E. Capacity of water cooled condenser can be determined from water flow and water temperature rise from standard formulas. Compare with system cooling coil capacity but add compressor motor input.

F. Refrigerant readings at evaporator coil should be made if piping between evaporator coil and compressor is long or if the compressor is several feet above the coil. (SP1 – SP2) indicates refrigerant suction pressure drop which should not be more than 5 PSI. (ST2 – ST1) indicates refrigerant suction temperature rise which should not be more than 3 degrees. Check pipe insulation. (ST2) should be about 5 degrees warmer than the equivalent temperature of (SP2). If colder, adjust expansion valve. (See System Analysis of Cooling Coil.)

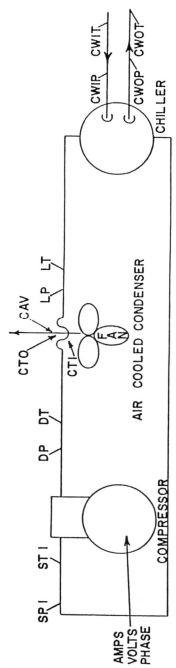

Figure 3-3   R3 / Direct Expansion Refrigeration / Air Cooled Chiller
Field Measurements

R3
DIRECT EXPANSION REFRIGERATION AIR COOLED CHILLER
FIELD MEASUREMENTS

SP1	Refrigerant Suction Pressure at Compressor	SP1 _____
ST1	Refrigerant Suction Temperature at Compressor	ST1 _____
DP	Refrigerant Discharge Pressure at Compressor	DP _____
DT	Refrigerant Discharge Temperature at Compressor	DT _____
LP	Refrigerant Liquid Pressure	LP _____
LT	Refrigerant Liquid Temperature	LT _____
CTI	Condenser Air Temperature In	CTI _____
CTO	Condenser Air Temperature Out	CTO _____
CAV	Condenser Air Volume	CAV _____
CWIP	Chilled Water Pressure In	CWIP _____
CWIT	Chilled Water Temperature In	CWIT _____
CWOP	Chilled Water Pressure Out	CWOP _____
CWOT	Chilled Water Temperature Out	CWOT _____
AMPS	Compressor Motor	AMPS _____
Phase	Compressor Motor	Volts _____
		Phase _____
		Type of Refrigerant _____

R3
REFRIGERATION
DIRECT EXPANSION AIR COOLED CHILLER

A. Compressor suction and discharge pressures are a function of the operation and determine the power input to the compressor motor. The lower the discharge pressure (DP) and the higher the suction pressure (SP1), the less KW the motor requires.

B. Suction temperature (ST1) should be higher than the equivalent refrigerant pressure-temperature so that liquid refrigerant does not return to compressor and cause damage.

C. Discharge gas temperature (DT) is also higher than the equivalent refrigerant pressure temperature because motor heat is added. If (DT) is not above equivalent temperature, refrigerant system needs examination for refrigerant volume, oil, air.

D. The air cooled condenser changes the refrigerant hot gas to liquid by removing the latent heat of vaporization. Liquid pressure (LP) should be very close to discharge gas temperature. Liquid temperature (LT) should be lower than discharge temperature (DT).

E. (a) Capacity of air cooled condenser is determined by the fan air volume (CAV), the entering air temperature (CTI) and leaving air temperature (CTO).

   (b) If air quantity is low, the coil is dirty, or there is excessive oil in the condenser coil the capacity may be low. Compare BTU/hr. condenser capacity with system cooling capacity plus compressor motor input. See standard formulas.

F. (a) Condition of the liquid chiller is critical not only from the energy standpoint but from the possibility of freeze-up and damage. Usual water temperature drop (CWIT − CWOT) is 8 degrees to 10 degrees F. Chilled water outlet temperature should not exceed 4 degrees above refrigerant suction temperature (STI). Unless chilled water has antifreeze, refrigerant temperature (STI) should not be below 36 degrees F.

(b) Chilled water pressure drop (CWIP – Crop) varies whether the water is in tubes or shell. Pressure drops usually vary between 5 and 25 PSI.

(c) Too low a pressure drop may indicate too low a water flow which can cause freeze-up.

(d) If chilled water leaving temperature (CWOT) varies more than 5 degrees with inlet temperature difference (CWIT – CWOT) remaining constant, check compressor size against cooling load. Compressor may be oversized for cooling load.

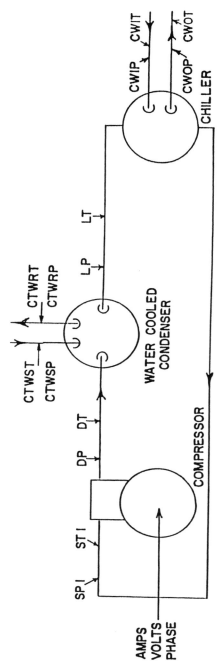

Figure 3-4 R4 / Direct Expansion Refrigeration / Water Cooled Chiller
Field Measurements

R4
DIRECT EXPANSION REFRIGERATION
WATER COOLED CHILLER
FIELD MEASUREMENTS

SP1   Refrigerant Suction Pressure at
       Compressor

ST1   Refrigerant Suction Temperature at
       Compressor

DP   Refrigerant Discharge Pressure at
       Compressor

DT   Refrigerant Discharge Temperature
       at Compressor

LP   Refrigerant Liquid Pressure

LT   Refrigerant Liquid Temperature

CTWSP Cooling Tower Water Supply
       Pressure

CTWST Cooling Tower Water Supply
       Temperature

CTWRP Cooling Tower Water Return
       Pressure

CTWRT Cooling Tower Water Return
       Temperature

CWIP   Chilled Water Pressure In

CWIT   Chilled Water Temperature In

CWOP   Chilled Water Pressure Out

CWOT   Chilled Water Temperature Out

AMPS   Compressor Motor

Volts Compressor Motor

Phase Compressor Motor

(Well water may be substituted for cooling
tower water.)

SP1 _____
ST1 _____
DP _____
DT _____
LP _____
LT _____
CTWSP_____
CTWST_____
CTWRP_____
CTWRT_____
CWIP _____
CWIT _____
CWOP _____
CWOT _____
AMPS_____
Volts _____
Phase_____
Type of
Refrigerant _____

R4
REFRIGERATION
DIRECT EXPANSION WATER COOLED CHILLER

A. Compressor suction and discharge pressures are a function of the operation and determine the power input to the compressor motor. The lower the discharge pressure (DP) and the higher the suction pressure (SP1), the less KW the motor requires.

B. Suction temperature (ST1) should be higher than the equivalent refrigerant pressure-temperature so that liquid refrigerant does not return to compressor and cause damage.

C. Discharge gas temperature (DT) is also higher than the equivalent refrigerant pressure temperature because motor heat is added. If (DT) is not above equivalent temperature, refrigerant system needs examination for refrigerant volume, oil, air.

D. (a) The water cooled condenser changes the refrigerant discharge gas to liquid by removing the latent heat of vaporization. Refrigerant liquid pressure (LP) should be slightly less than discharge pressure (DP). If this pressure drop (DP − LP) is over 5 PSI, check condition of condenser (refrigerant is usually in the shell).

   (b) Condenser cooling water either from cooling tower, wells or city water, usually has a temperature rise (CTWST − CTWRT) of 10 degrees F at full capacity for cooling towers, 95 degrees in, 85 degrees out maximum; for well or city water 20 degrees F rise if water supply is restrictive or expensive. The lower the return water temperature, the lower the refrigerant discharge pressure, the lower the compressor power input.

   (c) Condenser water pressure drop usually varies around 15 PSI. Higher pressure drop (CTWSP − CTWRP) may indicate restrictions in water tubes.

   (d) Low condenser water temperature rise (CTWST − CTWRT) plus low refrigerant temperature drop (DT − LT) may indicate low refrigerant volume, air in system, excessive oil.

E. Capacity of water cooled condenser can be determined from water flow and water temperature rise from standard formulas. Compare with system cooling coil capacity but add compressor motor input.

F. (a) Condition of the liquid chiller is critical not only from the energy standpoint but from the possibility of freeze-up and damage. Usual water temperature drop (CWIT − CWOT) is 8 degrees to 10 degrees F. Chilled water outlet temperature should not exceed 4 degrees above refrigerant suction temperature (ST1). Unless chilled water has antifreeze, refrigerant temperature (STI) should not be below 36 degrees F.

(b) Chilled water pressure drop (CWIP − CWOP) varies whether the water is in tubes or shell. Pressure drops usually vary between 5 and 25 PSI.

(c) Too low a pressure drop may indicate too low a water flow which can cause freeze-up.

(d) If chilled water leaving temperature (CWOT) varies more than 5 degrees with inlet temperature difference (CWIT − CWOT) remaining constant, check compressor size against cooling load. Compressor may be oversized for cooling load.

(e) See standard formula for chiller cooling capacity.

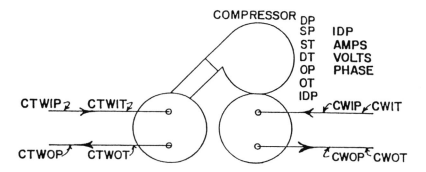

Figure 3-5   R5 / Refrigeration / Centrifugal Compressor
Field Measurements

R5
REFRIGERATION CENTRIFUGAL COMPRESSOR
FIELD MEASUREMENTS

SP	Refrigerant Suction Pressure	SP	_____
ST	Refrigerant Suction Temperature	ST	_____
DP	Refrigerant Discharge Pressure	DP	_____
DT	Refrigerant Discharge Temperature	DT	_____
OP	Refrigerant Oil Pressure	OP	_____
OT	Refrigerant Oil Temperature	OT	_____
IDP	Compressor Inlet Damper Position	IDP	_____
AMPS	Compressor Motor	AMPS	_____
Volts	Compressor Motor	Volts	_____
Phase	Compressor Motor	Phase	_____
CTWIP	Cooling Tower Water Inlet Temperature	CTWIP	_____
CTWIT	Cooling Tower Water Inlet Pressure	CTWIT	_____
CTWOP	Cooling Tower Water Outlet Pressure	CTWOP	_____
CTWOT	Cooling Tower Water Outlet Temperature	CTWOT	_____
CWIP	Chilled Water Inlet Pressure	CWIP	_____
CWIT	Chilled Water Inlet Temperature	CWIT	_____
CWOP	Chilled Water Outlet Pressure	CWOP	_____
CWOT	Chilled Water Outlet Temperature	CWOT	_____

Type of
Refrigerant _____

R5
REFRIGERATION
CENTRIFUGAL COMPRESSOR

A. Compressor suction and discharge pressures are a function of the operation and determine the power input to the compressor motor. The lower the discharge pressure (DP) and the higher the suction pressure (SP), the less KW the motor requires. Note that different refrigerants are used so pressures may vary. Refer to tables for refrigerant used.

B. Suction temperature (ST) should be higher than the equivalent refrigerant pressure-temperature so that liquid refrigerant does not return to compressor and cause damage. This is not a major problem on these machines.

C. Discharge gas temperature (DT) is also higher than the equivalent refrigerant pressure temperature because motor heat is added. If (DT) is not above equivalent temperature, refrigerant system needs examination for refrigerant volume, oil, air.

D. (a) The water cooled condenser changes the refrigerant discharge gas to liquid by removing the latent heat of vaporization.

   (b) Condenser cooling water either from cooling tower, wells or city water, usually has a temperature rise (CTWOT – CTWIT) of 10 degrees F at full capacity for cooling towers, 95 degrees in, 85 degrees out maximum; for well or city water 20 degrees F rise if water supply is restrictive or expensive. The lower the return water temperature, the lower the refrigerant discharge pressure, the lower the compressor power input.

   (c) Condenser water pressure drop usually varies around 15 PSI. Higher pressure drop (CTWIP – CTWOP) may indicate restrictions in water tubes.

   (d) Low condenser water temperature rise (CTWOT – CTWIT) plus low refrigerant temperature drop (DT – LT) may indicate low refrigerant volume, air in system, excessive oil.

E. Capacity of water cooled condenser can be determined from water flow and water temperature rise from standard formulas. Compare with system cooling coil capacity but add compressor motor input.

F. (a) Condition of the liquid chiller is critical not only from the energy standpoint but from the possibility of freeze-up and damage. Usual water temperature drop (CWIT – CWOT) is 8 degrees to 10 degrees F. Chilled water outlet temperature should not exceed 4 degrees above refrigerant suction temperature (ST). Unless chilled water has antifreeze, refrigerant temperature (ST) should not be below 36 degrees F.

    (b) Chilled water pressure drop (CWIP – CWOP) varies whether the water is in tubes or shell. Pressure drops usually vary between 5 and 2S PSI.

    (c) Too low a pressure drop may indicate too low a water flow which can cause freeze-up.

    (d) If chilled water leaving temperature (CWOT) varies more than 5 degrees with inlet temperature difference (CWIT – CLOT) remaining constant, check compressor size against cooling load. Compressor may be oversized for cooling load.

    (e) See standard formula for chiller cooling capacity.

G. Centrifugal operating characteristics.

    (a) Centrifugal compressors using some refrigerants operate under vacuum on suction side. It is mandatory that purge unit operate to remove air which may leak into the system,

    (b) Refer to manufacturer's instructions for your specific machine.

    (c) DO NOT CHANGE PHASES ON MOTOR. Reversing motor may loosen impellers and damage compressor.

    (d) At full load, centrifugal compressors require .8 to .9 KW per ton of refrigeration developed.

    (e) Minimum operating load is approximately 40%. Do not cycle machine for short time periods.

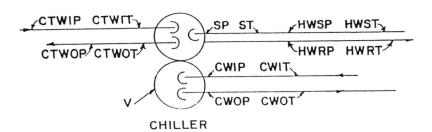

Figure 3-6 R6 / Absorption Chiller
Field Measurements

R6
ABSORPTION CHILLER
FIELD MEASUREMENTS

SP =	Steam Pressure	SP _____
ST =	Steam Temperature	ST _____
(OR)		HWSP_____
HWSP	Hot Water Supply Pressure	HWST_____
HWST	Hot Water Supply Temperature	HWRP_____
HWRP	Hot Water Return Pressure	HWRT_____
HWRT	Hot Water Return Temperature	CTWIP _____
CTWIP	Cooling Tower Water Inlet Temperature	CTWIT _____
CTWIT	Cooling Tower Water Inlet Pressure	CTWOP_____
CTWOP	Cooling Tower Water Outlet Pressure	CTWOT_____
CTWOT	Cooling Tower Water Outlet Temperature	CWIP _____
CWIP	Chilled Water Inlet Pressure	CWIT _____
CWIT	Chilled Water Inlet Temperature	CWOP_____
CWOP	Chilled Water Outlet Pressure	CWOT_____
CWOT	Chilled Water Outlet Temperature	V _____
V =	Operating Vacuum Inches Mercury	

R6
REFRIGERATION
ABSORPTION CHILLER

A. The absorption refrigeration system operates bv means of a chemical which cools off when it absorbs water, called the "heat of solution". The chemical usually used is lithium bromide. In order to
make the system work, the dilute lithium bromide is then heated by
means of hot water or steam, the water is evaporated from the
chemical, is separated, the chemical is cooled and returned to the
absorption chiller and the water is returned for solution again.

B. The absorption chiller, therefore, does not have an electric motor
other than a pump to circulate the refrigerant and is powered by
means of steam, preferably at 12 lbs. per square inch pressure but
may operate down to 8 lbs., or by hot water operated between 260
degrees or lower. Some machines operate as low as 170 degree
water which operates from solar panels but is rather inefficient and
requires a very large machine for capacity.

C. Air in the refrigerant system causes difficulties not only in the
capacity of system but since the lithium bromide is corrosive, the
presence of air will cause considerable damage to the machine over
a period of time. There is a mercury manometer on the side of the
machine which indicates the vacuum under which the machine
operates and this is a very important operating condition. Check the
operating manual for the proper vacuum which is about 4 inches of
mercury.

D. The amount of heat necessary to create a ton of cooling varies on
the new machines at about 14.5 lbs. of steam per ton of refrigeration to the older machines at approximately 19 lbs. per ton.

E. Because of the heat added by the steam or hot water, the condenser
water to the cooling tower (CTWIT) is in the range of 101.5
degrees and the cooling tower water returning to the chiller
(CTWOT) is normally at 85 degrees. A by-pass control or thermostat control must be used on these chillers as at condenser water
temperatures below 54 degrees the refrigerant crystallizes out of the
water and solidifies in the machine. This may require taking the

machine apart, entering same and breaking this material loose with a sledge hammer. This minimum condenser temperature is very critical on these machines.

F.   Chilled water inlet and outlet temperatures are similar to other machines operating in the range of 40 to 50 degrees. The higher the water temperature, the more efficient the refrigeration equipment will operate. The lower the condenser water temperature within the safe limits, the higher the capacity of the refrigeration machine.

G.   The absorption refrigeration machine has a peculiar requirement that if not properly controlled, the amount of steam or hot water entering the machine on start-up is approximately five times that required for normal continuous operation. If this high surge of heat supply causes difficulty in the rest of the system, it is recommended that the rate of buildup of capacity on the machine be extended to approximately 20 minutes and the rate of steam or hot water admitted is slowed down to approximately 105 above normal operation. This can be performed by a restriction in the control line to the steam valve or the control device if it is pneumatic with provisions to allow the control to reverse without restriction.

H.   Absorption refrigeration machines can operate satisfactorily down to about 11% capacity with an input of about 22% of total energy consumption. This provides a more efficient operation at low capacities than the centrifugal refrigeration unit.

I.   These chillers are also susceptible to damage if the water flow through the chiller or condenser is too low for the minimum capacity of the machine. Check the manufacturer's manual for all characteristics of these machines as they will vary in accordance with the size, construction and make of absorption chiller.

J.   Do not operate this equipment at steam pressures or water temperatures higher than that satisfactory for the particular device being used.

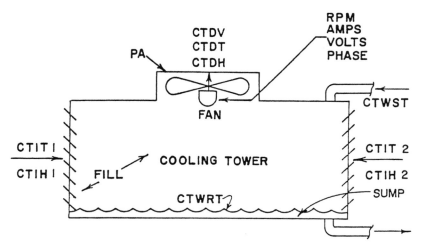

Figure 3-7 R7 / Cooling Tower
Field Measurements

R7
COOLING TOWER
FIELD MEASUREMENTS

CTDV	Cooling Tower Air Discharge Volume	CTDV _____
CTDT	Cooling Tower Air Discharge Temperature	CTDT _____
DTDH	Cooling Tower Air Discharge Wet Bulb Temperature	CTDH _____
		CTIT1 _____
CTIT1, 2	Cooling Tower Air Inlet Temperature	CTIT2 _____
CTIH1, 2	Cooling Tower Air Inlet Wet Bulb Temperature	CTIH1 _____
		CTIH2 _____
CTWST	Cooling Tower Water Supply Temperature	CTWST _____
CTWRT	Cooling Tower Water Return Temperature	CTWRT _____
RPM	Fan Speed	RPM _____
AMPS	Fan Motor	Amps _____
Volts	Fan Motor	Volts _____
Phase	Fan Motor	Phase _____
PA	Fan Blade Pitch Angle	PA _____
Fill	Type of inside material – Wood, Plastic, Metal, Nothing?.	Fill _____

R7
REFRIGERATION
COOLING TOWER

A. Cooling towers are of many configurations. They all provide water cooling by partial evaporation of water into the air. This increases the temperature and humidity of air through the tower.

B. Cooling towers are normally rated at supply water (CTWST) at 95 degrees for electric and 101.5 degrees for absorption refrigeration. Return water (CTWRT) is usually at 85 degrees or within 7 to 8 degrees of entering wet bulb temperature.

C. Cooling tower capacity is determined by fan discharge volume (CTDV), discharge wet bulb temperature (CTDH) and inlet air wet bulb temperature (CTIH) (average CTIH1 and CTIH2 if tower has more than one inlet). Wind velocity from one side affects entering air conditions so average readings are necessary.

D. Since increasing the cooling tower capacity produces a lower return water temperature and therefore more efficient refrigeration, determine if the cooling tower is at peak performance. Limitation – absorption refrigeration requires a minimum water temperature.

Check amp reading of fan motor against motor rating. If operating amps is below motor rating, fan capacity can be increased.

E. If fan drive is belt driven, change motor sheave. If direct drive with variable pitch fan blades, increase pitch.

F. Limitations: If fan discharge temperature (CTDT) and wet bulb (CTDH) is within one degree and there are water droplets in discharge air, increased air flow will cause increased water carry-over.

G. Fill or inside material of tower affects the contact of water and air. Minerals, too smooth surface or restricted air flow will reduce capacity. Algae growth affects tower surfaces. It also breaks down and travels with the water, plugging up heat exchangers.

DO NOT PUT GLYCOL ANTI-FREEZE IN WOOD OR PLASTIC FILL TOWERS. PERMANENT DAMAGE RESULTS.

H. Examine water flow out of the sump. Vortexing or swirling at open-
    ing picks up air bubbles reducing cooling effect of water and may
    damage pumps. Eliminate vortexing by putting a grating over opening.

I. Sump water level should be not less than 5. if operating above outside
    temperatures above 32 degrees F. Lower levels increase vortexing.

J. If tower has more than one motor driven fan, consider operating fans
    in steps from water return temperature (CTWRT) in sump. Two speed
    fans controlled in same manner, even if only one is on tower, will also
    reduce operating power.

TYPICAL AIR COOLED CHILLERS – 1980 MODEL
90° AIR, 44° WATER

Actual Ton Capacity	CFM Fan	Compressor KW	Fans	CFM/Ton
18	18,200	18.2	3-26"	1,000
26	28,200	27.0	3-30"	1,084
35.2	26,000	39.0	4-26"	739
51	39,000	56.5	6-26"	806
67	54,000	74.7	6-30"	806
80.5	79,200	93.5	2-84"	984
100	75,000	112	2-84"	750
117	82,800	163	2-84"	708

CONDENSER FACE VELOCITY VARIES FROM 350 TO 550 FT. PER MINUTE

Mfg. No. 2	CFM	CFM/Ton
19.3	15,810	819
29.6	26,100	882
54.5	53,000	972
71.4	114,000	1,597
100	110,000	1,100
120	100,200	835

TYPICAL COOLING TOWER CAPACITY

Rated Tons	CFM Fans	CFM/Ton
10	3,800	380
20	4,630	230
40	9,500	237
65	15,300	235
75	17,000	226
105	23,100	226
150	33,460	223
215	49,770	231
265	51,400	194
395	76,600	194
510	98,400	193

CHILLED WATER COIL CAPACITY
80° DB, 68° WB ENTERING
42° ENTERING WATER, 10° RISE

Fins	Rows	Face Velocity Ft/Min	MBH	LVG °D.B.	LVG °W.B.
8	4	400	16.4	56.5	55
		500	18.6	58.3	56.4
		600	20.3	59.7	57.5
		700	21.9	60.9	58.4
14	4	400	19.5	52.5	52.5
		500	22.1	54.5	53.8
		600	24.3	56.0	55.2
		700	26.2	57.4	56.3
14	6	400	23.4	48.4	48.3
		500	27.3	50	49.9
		600	30.6	51.5	51.3
		700	33.5	52.8	52.5
14	8	400	25.7	46	45.7
		500	30.5	47.3	47.2
		600	34.7	48.6	48.5
		700	38.6	49.7	49.6

Chapter 4

Computer Analysis

(See Diagrams)

TEMPERATURE

T-1 *Room Air Temperature* – if temperature is too high check setting of sensor. Check supply air temperature to be at normal setting, roughly 55°. If above this point, check further operations conditions. If the supply temperature is correct call the occupant in the space and determine if there is any sun shining on the thermostat or there is any heat produced surrounding the thermostat. If temperature is higher in other parts of the room away from the thermostat, this requires a test of the air circulating system to see if the cool air is distributing throughout the space.

T-2 *Room Temperature Low* – Check sensor setting.

T-2 *Supply Air Temperature* – If temperature is above normal setting check to see if compressor is running which can be determined by T-7 which is the discharge air off the air conditioning condenser which will indicate whether the compressor is running.

T-2 *Supply Air Temperature Low* – Determine if the total air supply V-1 is correct and if not, further investigation will have to be made to bring the supply volume up to normal. Also check the air volume capacity from the static pressure, S-2 minus S-1 which will give the static pressure across the air conditioning unit to determine whether there is any problem with dirty coils or restriction in the air conditioning unit itself.

T-3 *Mixed Air Temperature* – This is variable according to the temperature of the outside air and the return air. You can use the formula
in the Handbook to determine the percentage of outside air coming
in. Since about 20 CFM per person is required, you can calculate
the requirements of outside air and reduce the same by changing
the position of tile damper motors DM-1 and DM-2 to control the
amount of outside air being brought into the unit.

T-4 *Outside Air* – A measurement of the outside air conditions which
will determine whether the system should be on cooling or heating. Also, if it is within a range of temperature and the outside air
humidity is not too high, you can use this outside air to cool with
without operating the refrigeration system.

T-5 *Return Air* – This should be within 2° of the average room temperature in the apace. If it is much below this, and the system is
not a bypass VAV system. There is a possibility of duct leaks
especially in those places where the return space above a ceiling is
used for return air.

T-6 The air entering an air cooled condenser should be the same as
outside air T-4. If it is above this temperature, it is recirculating
causing a decrease in capacity and efficiency and should be examined for what is causing this and method of prevention.

T-7 *Air Discharge* – Condenser air temperature should be approximately 20° above the entering air temperature at full capacity of
compressor. If the temperature rises above this amount it is possible that there may be air or moisture in the refrigerant circuit and
the system should be checked. If the temperature difference is less
than 20°, something may be wrong with the refrigeration operation
as it may not be supplying full capacity.

HUMIDITY

If you have humidity sensing instruments on the system, you check
the H-1 supply air humidity which should be pretty close to 100%, if
the coils are doing a reasonably good dehumidifying job.

H-2 *Mixed Air Humidity* – Is primarily the result of the moisture in the
outside air mixing with the moisture in the return air and in most
cases, it is above the room humidity if the outside air has high
moisture content. This is particularly true at night time in the

Summer in high humidity areas such as the Southern United States and Gulf Coast and coastal areas.

H-3 *Room Air Humidity* – It is important as this should be kept preferably between 40 and 60% for most comfort and health reasons. If it exceeds 70%, you may get respiratory problems and if it is below 30% you also may get respiratory problems of a reverse kind.

STATIC PRESSURE

S-1 *Fan Inlet Pressure* – That gives you the static pressure required to overcome the resistance of the coils, filters, and return air system. Increasing this negative pressure will indicate dirty coils and filters or some restrictions in the return ducts.

S-2 *Supply Air Pressure* – Gives you the conditions of the duct work and any air outlets which may encounter some resistance due to shift in duct conditions or dirt in the system, or whether the lining came loose and blocked off part of the turning vanes or dampers.

S-1 minus S-3 will give you the pressure drop across the coils, which will determine whether the coils are functioning and allowing sufficient air to flow through it. The difference between S-3 and S-4 will give you the pressure drop across the filter to determine if the filter needs replacing.

In general, on static pressures, if there is a low supply air volume, V-1, and a high static pressure difference, S-1 minus S-3, it may be advisable to check the fan blades especially if it is a forward curved blade fan to find out if there is dirt in the blades. Refer back to Page 209, you will see the effect of dirt on these blades.

AIR VOLUME

If you have volume indicating instruments, a reduction in the supply air volume indicates abnormal resistance in the air conditioning unit coils and filters, or possibly slipping of the belts or dirt in the fan. If the supply air temperature has dropped below 55°, then the air volume is low. It may result in mold generation on the air outlets because of low temperatures. This is particularly true in high humidity areas and may not be of too much concern in dry areas where it is possible to supply air as low as 45° without difficulty.

WATER COOLED SYSTEMS

This particularly applies to individual air conditioning units supplied with condenser water from a cooling tower or heat exchanger and is popular in apartment houses, condominium units, and office buildings as well as heat pump type units for summer and winter operation.

T-1 *Room Temperature* – This is a similar condition as the previous section. If the temperature is too high, again examine the supply air temperature T-2 and also the setting the room is at and the conditions that the thermostat might be exposed to.

T-2 *Room Supply Temperature* – Again, it is important to determine the operation of equipment. If the temperature is too high, examine the possibility that the compressor may not be operating properly and this can be determined by the difference between the entering water temperature T-6, and the leaving water temperature T-3, to determine if there is the same normal rise of about 8° or whatever was originally set up for that particular unit. A decrease in the difference between these temperatures may indicate that the compressor is not working full capacity or that the condenser coil may be plugged up with lime in the tubes.

In operation of the cooling tower, T-4 should be the normal outside air temperature. If it is higher, it is possible that there may be some recirculation of air from the discharge of the cooling tower T-5 and the difference between T-4 and T-5 should be approximately 8 to 10°, depending on the original capacity of the cooling tower. The water supply temperature, T-6, should be roughly 85° with the return temperature maximum of 95° during maximum outside air conditions. If these temperatures can be lowered, the efficiency of the air conditioning unit goes up as less KW is required at lower condenser temperatures. There are some limitations on some of this equipment as there are certain types of refrigeration systems that are susceptible to low condenser water temperatures and should be checked.

On a condenser water system, the most important conditions are the suction and discharge pressure at the pumps. In a number of buildings, if the pumps and the cooling tower are on top of a building more than three stories, quite often the city pressure may not be sufficient to get the water up to the top of the tower in full volume so that the suction pressure, P-1 and discharge pressure, P-2 of the pump is very important. The suction pressure on the pump should be not less than 4 pounds per square inch if it is a 1750 RPM pump, and roughly 10 pounds per square inch if it is a 3500 RPM pump or the pump may not operate at full capacity. Also, if there is a low suction pressure and there is a strainer between the pump and the cooling tower base, the strainer may be plugged up and should be investigated. Discharge pressure, P-2, should be sufficient to distribute the water throughout the system and put the return water back to the top of the tower either into the spray or the distribution water trough in sufficient capacity to show complete distribution of water throughout the tower. In some all-glass office buildings which have small computer rooms which require cooling 24 hours a day, 7 days a week, it may be possible to so condition these areas even though the system is shut down on nights and weekends. If a check valve is installed in the discharge line from the condenser on each unit, it is possible to put a circulating pump on those units for the computer rooms only and just supply sufficient water to take care of those areas and put the cooling tower fan on a water temperature thermostat to operate the fan when required.

In one case, in a 100,000 square foot office building, it was possible to reduce the requirements for pumping in the building from 20 horsepower down to 1/3 horsepower due to this type of control during the shut-down periods of nights and weekends.

The checking of the temperature difference between T-6 and T-3 which indicates the temperature rise through the water cooled condenser on the unit itself is extremely important because of lime generation due to poor water treatment. In one case, in a condominium that was using well water for cooling, the double tube condenser on the small residential type air conditioners had to be replaced twice a year because they plugged up.

If these units are heat pump type, there should be an indicator in the system showing whether the control valve is set for heating or for cooling which reverses the operation of the refrigeration circuit. Operation of the air conditioning unit itself from the standpoint of air volume, supplied temperature, room temperature, static pressure should be similar to previous discussion on single-zone air conditioning systems. The "on", "off" indicator for the individual air conditioning unit fans, the pump, and the cooling tower fan are recommended as an indication of whether this equipment is functioning properly or not. If the computer screen has sufficient facilities, it is possible that you may be able to put the basic setting data and the instruction data on that same screen or on an additional screen such as shown on page 99.

If the complaint is from an area and a person who is not satisfied, refer to page 200 which indicates a number of the problems that may be affecting that person's comfort level which may or may not be temperature or humidity, but may be stress or physical conditions.

On the heating cycle with heat pumps, investigation should be made to determine if the system has switched over to heating. If there are electrical heating elements in the unit, whether those units are on, and especially take a look at the condition of the quantity of outside air which has a very strong effect on the requirements for heating.

If steam or hot water is used for heating in a coil, sensing devices determining the position of the control valves would be of value to determine if the control equipment is supplying sufficient heating medium to the coils.

During heating, the air distribution is also a very important consideration as the warm air has a tendency to rise towards the ceiling, and if there is insufficient velocity, the air may not get down to the breathing level or the occupied level. If there are ceiling fans, it is advisable to operate them in reverse which will remove the heat from the ceiling and put it down towards the floor. On central air conditioning systems, the zone reheating coils used either for control or for dehumidification, inspection should be made on these coils as to the operating condition referring back to the section on A Hospital Test, page 225 which indicated the problems that may arise due to heating conditions. Here, again, the problems encountered under page 200 should also be considered for winter time operation, too.

Additional controls for minimum cooling tower or condenser water systems for centrifugal and absorption machines may be required to prevent too cold water entering this equipment and causing damage.

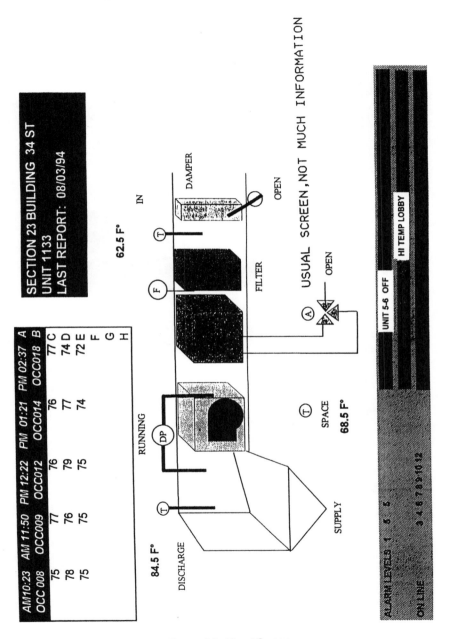

Figure 4-1 Usual Screen

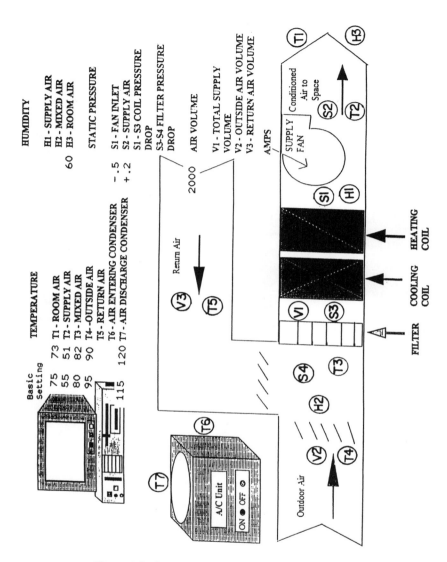

Figure 4-2 Computer Control of an Air Conditioning System

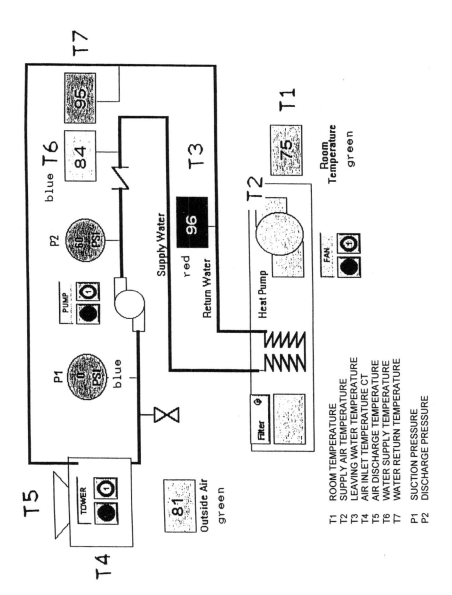

Figure 4-3 Heat Pump

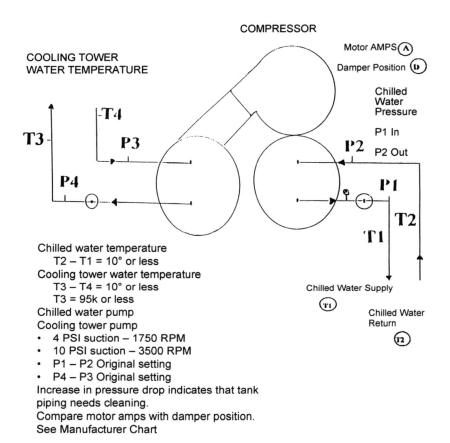

Chilled water temperature
   T2 – T1 = 10° or less
Cooling tower water temperature
   T3 – T4 = 10° or less
   T3 = 95k or less
Chilled water pump
Cooling tower pump
• 4 PSI suction – 1750 RPM
• 10 PSI suction – 3500 RPM
• P1 – P2 Original setting
• P4 – P3 Original setting
Increase in pressure drop indicates that tank
piping needs cleaning.
Compare motor amps with damper position.
See Manufacturer Chart

Figure 4-4 Centrifugal Compressor

SYSTEM ANALYSIS

Do this before calling maintenance:

1. Get call-room warm

2. Check room temperature (T1)

3. Check room humidity (H1)

4. If ok, check supply air temperature (T2)

5. If (T2) is below 55°, check (V1) for sufficient volume.

7. If (T2) is above 60°, is compressor on?

8. If compressor on check, (T6) & (T7) , should be 20° difference.

   Is (T6) the same as (T4) ? If (T6) is higher, air is recirculating.

   If (T7) is low, compressor is not working.

9. Check (V2) and (V3) for correct percent of outside air.

   (T3) (T4) (T5) calculation can also show percent of outside air.

10. If (T5) is lower (T1) , ducts are leaking.

11. (S2) – (S3) , should be at initial pressure difference.

12. If (S2) is high, there may be an obstruction in duct or Closed damper.

13. If (S1) – (S3) is high, coils are dirty.

14. If (S2) – (S1) is low, fan is dirty or fan belt is slipping.

15. If (S3) – (S4) is high, filter is dirty.

16. Now if everything checks out and you are still warm go look at room sensor and see if the sun is shining on it.

17. Now you have complete information on the system and can tell maintenance exactly what to look for - saving about 75% of maintenance time.

Figure 4-5 Existing energy management system not used control replaced with remote
computer night and weekend shut down and step start up program.
Savings $5000 per month.

Chapter 5

Basic Formulas

AIR COOLING SYSTEMS

Cooling – sensible heat removal

CFM x (mixed air temperature – supply temperature) x 1.08 =
sensible heat removed BTU/hr.

Cooling – total heat removal

CFM x (wet bulb total heat [BTU/#] mixed air –
wet bulb total heat [BTU/#] supply air x

$$\frac{60}{14} = 4.3 = \text{total BTU per hour cooling}$$

CFM x (TH in – TH out) x 4.3 = total BTU per hour cooling.

(See table of total heat for each wet bulb temperature.)

REFRIGERATION PLANTS

Chilled Water Flow

GPM x 500 x (water temperature out – water temperature in) =
BTU per hour total cooling capacity.
GPM x 500 x (T out – T in) = BTU per hour.

Refrigerant capacity – 40k refrigerant

Water cooled   1 KW   = 1 ton cooling (12,000 BTU)
                     1 KWH = 12,000 BTU/hr.

Air cooled   1.1-1.4 KW   = 1 ton cooling (12,000 BTU)
                     1.1-1.4 KWH = 12,000 BTU/hr.

BASIC FORMULAS

CFM = cubic feet per minute
GPM = gallons per minute
  T = temperature kF
 TH = total heat BTU/#
KWH = kilowatt hours

1. *Air cooling systems*

   Cooling-sensible heat removal

   CFM x (T in -T out) x 1.08 = BTU per hour

   Cooling – total heat removal

   CFM x (TH in -TH out) x 4.3 = BTU per hour

2. *Refrigerant plants*

   A. Chilled water flow
       GPM x 500 x (T out -T in) = BTU per hour

   B. Electric refrigeration
       Water cooled   1 KWH = 12,000 BTU/hr. cooling
       Air cooled 1.1-1.4 KWH = 12,000 BTU/hr. cooling

   C. Absorption refrigeration
       19 pound steam per hour = 12,000 BTU per hour cooling

   D. Condensing systems
       1) Air cooled
           Operating capacity CFM x (T out -T in) x 1.08 = BTU per hour

           $\dfrac{\text{BTU/hr. capacity}}{16,098}$ = tons (12,000 BTU) per hour

2) Water cooled
   Cooling tower capacity
   CFM x (TH out -TH in) x 4.3 = BTU/hr. total capacity
   Water flow

$$\frac{\text{BTU/Hr. capacity}}{500 \times (\text{T water in -T water out})} = \text{G.P.M.}$$

3) Net refrigeration effect
   Electric:

$$\frac{\text{BTU per hour capacity}}{15,415} = \text{tons}$$

Absorption:

$$\frac{\text{BTU per hour capacity}}{31,000} = \text{tons}$$

3. *Mixed air temperature calculation if air volumes are known:*

Mixed air temperature = % volume x temperature of outdoor air
+ % volume x temperature of return air.

Example: 4000 CFM of outdoor air at 25°F
           6000 CFM of return air at 75°F
           10000 CFM total air volume

(T) mixed = 4000/10,000 x 25° + 6000/10000 x 75° =
           10°   +   45°   = 55°F

4. *Outdoor air volume calculation if temperatures are known:*

Outdoor air volume = supply air temperature minus % return air volume times
temperature times total volume divided by outdoor air temperature.

Example: Supply air temperature 55°
           Supply air volume 10,000 CFM
           Return air temperature 75°
           Return air volume 6,000 CFM
           Outdoor air temperature 25°

Outdoor air volume

$$\text{CFM} \quad = \frac{55° - 6000/10000 \times 75°}{25°} \times 10{,}000 \text{ CFM}$$

$$\frac{55° - 45°}{25°} \times 10{,}000 = 4{,}000 \text{ CFM}$$

REFRIGERATION CONDENSING

Air cooled

Condensing capacity required per ton cooling.

Cooling	12,000 BTU/hr.

$$* \text{ Motor input average 1.2 KW} \times 3415 = \frac{4{,}098}{16{,}098\,\text{BTU/hr.}}$$

Condenser operation

CFM x (air temperature out – air temperature in) x 1.08 = BTU per hour

$$\frac{\text{BTU per hour}}{16{,}098} = \text{tons cooling operation}$$

Water cooled

Electric condensing capacity required per ton cooling.

Cooling	12,000 BTU/hr.
Motor 1 KW x 3415	3,415
	15,415 BTU/hr.

Steam absorption condensing capacity required per ton cooling.

Cooling	12,000 BTU/hr.
* Steam 19# x 1,000 BTU	19,000
	31,000 BTU/hr.

COOLING TOWER OPERATION

Operating capacity

Fan CFM x (wet bulb temperature total heat out
– wet bulb temperature total heat in) x 4.3 = BTU per hour

CFM x (BTU/# TH out -BTU/# TH in) x 4.3
= BTU per hour total operating capacity

Water flow

BTU per hour operating capacity ÷ 500
x (temperature of water in – temperature of water out) = GPM water flow

$$GPM = \frac{BTU\ per\ hour}{500 \times (T\ in\ -Tout)}$$

ELECTRIC FORMULAS

Single phase:

$$\frac{Amps \times volts}{1,000} = KW$$

Three phase:

$$\frac{Amps \times volts \times 1.732}{1,000} = KW$$

1 KWH = 3,412 BTU/hr.

REFRIGERANT TEMPERATURE-PRESSURE RELATION
Refrigerant Type 11

Temp. °F	Pressure PSIG	Temp. °F	Pressure PSIG	Temp. °F	Pressure PSIG	Temp. °F	Pressure PSIG	Temp. °F	Pressure PSIG
−60	29.1*	20	21.0*	65	5.3*	100	8.7	170	54.7
−55	29.0*	22	20.6*	66	4.8*	102	9.6	172	56.6
−50	28.8*	24	20.1*	67	4.3*	104	10.4	174	58.6
−45	28.6*	26	19.6*	68	3.8*	106	11.3	176	60.6
−40	28.4*	28	19.1*	69	3.2*	108	12.2	178	62.6
−38	28.3*	30	18.6*	70	2.7*	110	13.1	180	64.7
−36	28.1*	32	18.0*	71	2.2*	112	14.1	182	66.8
−34	28.0*	34	17.4*	72	1.6*	114	15.1	184	68.9
−32	27.9*	36	16.8*	73	1.0*	116	16.1	186	71.1
−30	27.8*	38	16.2*	74	0.5*	118	17.1	188	73.4
−23	27.6*	40	15.6*	75	0.0	120	18.2	190	75.7
−26	27.5*	41	15.2*	76	0.3	122	19.3	192	78.0
−24	27.3*	42	14.9*	77	0.6	124	20.4	194	80.4
−22	27.1*	43	14.6*	78	0.9	126	21.5	196	82.8
−20	27.0*	44	14.2*	79	1.2	128	22.7	198	85.3
−18	26.8*	45	13.9*	80	1.5	130	23.9	200	87.8
−16	26.5 *	46	13.5 *	81	1.8	132	25.2	210	101.1
−14	26.4*	47	13.1*	82	2.1	134	26.4	220	115.6
−12	26.2*	48	12.8*	83	2.4	136	27.7	230	131.4
−10	26.0*	49	12.4*	84	2.8	138	29.0	240	148.6
−8	25.7*	50	12.0*	85	3.1	140	30.4	250	167.3
−6	25.5*	51	11.6*	86	3.4	142	31.8	260	187.5
−4	25.2*	52	11.2*	87	3.8	144	33.2	270	209.3
−2	24.9*	53	10.8*	88	4.1	146	34.6	280	232.9
0	24.7*	54	10.4*	89	4.5	148	36.1	290	258.2
2	24.4*	55	9.9*	90	4.8	150	37.6	300	285.5
4	24.1*	56	9.5*	91	5.2	152	39.2	310	314.7
5	23.9*	57	9.1*	92	5.6	154	40.7	320	346.0
6	23.7*	58	8.6*	93	5.9	156	42.4	330	379.4
8	23.4*	59	8.2*	94	6.3	158	44.0	340	415.2
10	23.0*	60	7.7*	95	6.7	160	45.7	350	453.4
12	22.7*	61	7.3*	96	7.1	162	41.4	360	494.1
14	22.3*	62	6.8*	97	7.5	164	49.2	370	537.5
16	21.9*	63	6.3*	98	7.9	166	51.0	380	583.7
18	21.5*	64	5.8*	99	8.3	168	52.8		

(*) Vacuum = inches Mercury
     PSIG = pounds per square inch gauge

REFRIGERANT TEMPERATURE-PRESSURE RELATION
Refrigerant Type 12

Temp. °F	Pressure PSIG	Temp. °F	Pressure PSIG	Temp. °F	Pressure PSIG	Temp. °F	Pressure PSIG	Temp. °F	Pressure PSIG
-152	29.6*	-10	4.4	44	40.7	80	84.1	140	206.6
-150	29.6*	-8	5.3	45	41.6	81	85.6	145	220.3
-145	29.5*	-6	6.2	46	42.6	82	87.1	150	234.6
-140	29.3*	-4	7.1	47	43.6	83	88.6	155	249.5
-135	29.2*	-2	8.1	48	44.6	84	90.2	160	265.1
-130	29.0*	0	9.1	49	45.6	85	91.7	165	281.3
-125	28.8*	2	10.1	50	46.6	86	93.3	170	298.3
-120	28.6*	4	11.2	51	47.7	87	94.9	175	315.9
-115	28.3*	5	11.7	52	48.7	88	96.5	180	334.3
-110	27.9*	6	12.3	53	49.8	89	98.1	185	353.4
				54	50.9				
-105	27.5*	8	13.4			90	99.7	190	373.2
-100	27.0*	10	14.6	55	52.0	92	103.1	195	393.9
-95	26.4*	12	15.8	56	53.1	94	106.5	200	415.3
-90	25.7*	14	17.0	57	54.2	96	110.0	205	437.6
-85	24.9*	16	18.3	58	55.4	98	113.5	210	460.8
				59	56.5				
-80	24.0*	18	19.6			100	117.1	215	484.8
-75	23.0*	20	21.0	60	57.7	102	120.8	220	509.7
-70	21.8*	22	22.4	61	58.9	104	124.6	225	535.5
-65	20.5*	24	23.8	62	60.1	106	128.4	230	562.3
-60	19.0*	26	25.3	63	61.3	108	132.4		
				64	62.5				
-55	17.3*	28	26.8			110	136.4		
-50	15.4*	30	28.4	65	63.7	112	140.4		
-45	13.3*	31	29.2	66	65.0	114	144.6		
-40	10.9*	32	30.0	67	66.3	116	148.9		
-35	8.3*	33	30.8	68	67.5	118	153.2		
				69	68.8				
-30	5.4*	34	31.7			120	157.6		
-28	4.2*	35	32.5	70	70.1	122	162.1		
-26	2.9*	36	33.4	71	71.5	124	166.7		
-24	1.6*	37	34.2	72	72.8	126	171.4		
-22	0.2*	38	35.1	73	74.2	128	176.1		
				74	75.5				
-20	0.5	39	36.0			130	181.0		
-18	1.3	40	36.9	75	76.9	132	185.9		
-16	2.0	41	37.8	76	78.3	134	190.9		
-14	2.8	42	38.8	77	79.8	136	196.0		
-12	3.6	43	39.7	78	81.2	138	201.3		
				79	82.7				

(*) Vacuum – inches Mercury
PSI9 = pounds per square inch gauge

REFRIGERANT TEMPERATURE-PRESSURE RELATION
Refrigerant Type 22

Temp. °F	Pressure PSIG	Temp. °F	Pressure PSIG	Temp. °F	Pressure PSIG	Temp. °F	Pressure PSIG	Temp. °F	Pressure PSIG
−150	29.3*	−50	6.1*	18	40.8	88	163.2	158	419.8
−145	29.2*	−48	4.8*	20	43.0	90	168.4	160	429.8
−140	29.0*	−46	3.4*	22	45.2	92	173.6	162	440.0
−135	28.7*	−44	2.0*	24	47.5	94	179.0	164	450.3
−130	28.4*	−42	0.4*	26	49.9	96	184.5	166	460.9
−125	28.1*	−40	0.5	28	52.3	98	190.1	168	471.6
−120	27.6*	−38	1.3	30	54.8	100	195.9	170	482.5
−115	27.1*	−36	2.1	32	57.4	102	201.7	172	493.6
−110	26.5*	−34	3.0	34	60.1	104	207.7	174	504.9
−105	25.8*	−32	3.9	36	62.8	106	213.8	176	516.4
−100	25.0*	−30	4.8	38	65.6	108	220.0	178	528.1
−98	24.6*	−28	5.8	40	68.5	110	226.3	180	540.0
−96	24.2*	−26	6.8	42	71.4	112	232.8	182	552.2
−94	23.8*	−24	7.9	44	74.4	114	239.3	184	564.5
−92	23.4*	−22	9.0	46	77.5	116	246.1	186	577.1
−90	22.9*	−20	10.1	48	80.7	118	252.9	188	589.8
−88	22.4*	−18	11.3	50	84.0	120	259.9	190	602.8
−86	21.9*	−16	12.5	52	87.3	122	267.0	192	616.1
−84	21.3*	−14	13.8	54	90.8	124	274.2	194	629.6
−82	20.7*	−12	15.1	56	94.3	126	281.6	196	643.3
−80	20.1*	−10	16.4	58	97.9	128	289.1	198	657.3
−78	19.5*	−8	17.8	60	101.6	130	290.8	200	671.6
−76	18.8*	−6	19.3	62	105.3	132	304.6	202	686.2
−74	18.1*	−4	20.8	64	109.2	134	312.5	204	701.0
−72	17.3*	−2	22.3	66	113.2	136	320.6		
−70	16.5*	0	23.9	68	117.2	138	328.8		
−68	15.7*	2	25.6	70	121.4	140	337.2		
−66	14.8*	4	27.3	72	125.6	142	345.7		
−64	13.9*	5	28.1	74	130.0	144	354.4		
−62	12.9*	6	29.0	76	134.4	146	363.3		
−60	11.9*	8	30.8	78	138.9	148	372.3		
−58	10.9*	10	32.7	80	143.6	150	381.5		
−56	9.7*	12	34.7	82	148.3	152	390.8		
−54	8.6*	14	36.6	84	153.2	154	400.3		
−52	7.4*	16	38.7	86	158.1	156	409.9		

(*) Vacuum – inches Mercury
   PSIG – pounds per square inch gauge

REFRIGERANT TEMPERATURE-PRESSURE RELATION
Refrigerant Type 113

Temp. °F	Pressure PSIG	Temp. °F	Pressure PSIG	Temp. °F	Pressure PSIG
-30	29.2*	62	20.4*	135	5.3
-20	29.0*	64	20.0*	140	7.1
-10	28.6*	66	19.5*	145	9.0
0	28.1*	68	19.0*	150	11.0
2	28.0*	70	18.5*	160	15.5
4	27.9*	72	18.0*	170	20.5
5	27.8*	74	17.5*	180	26.1
6	27.8*	76	16.9*	190	32.4
8	27.6*	78	16.3*	200	39.4
22	27.5*	80	15.7*	210	47.1
12	27.3*	82	15.1*	220	55.6
14	27.2*	84	14.5*	230	65.0
16	27.0*	86	13.8*	240	75.3
18	26.9*	88	13.1*	250	86.5
20	26.7*	90	12.4*	260	98.7
22	26.5	82	11.7*	280	126.4
24	26.3*	94	10.9*	300	158.8
26	26.1*	96	10.1*	320	196.4
28	25.9*	98	9.3*	340	239.8
30	25.7*	100	8.5*	360	289.7
32	25.4*	102	7.6*	380	346.9
34	25.2*	104	6.7*	400	412.5
36	24.9*	106	5.8*	410	449.0
38	24.7*	108	4.9*	417	476.6
40	24.4*	110	3.9*		
42	24.1*	112	2.9*		
44	23.8*	114	1.9*		
46	23.5*	116	0.8*		
48	23.2*	118	0.1		
50	22.8*	120	0.6		
52	22.4*	122	1.2		
50	22.1*	124	1.8		
56	21.7*	126	2.4		
58	21.3*	128	3.0		
60	20.9*	130	3.6		

(*) Vacuum – inches Mercury
    PSIG – pounds per square inch gauge

REFRIGERANT TEMPERATURE-PRESSURE RELATION
Refrigerant Type 717 (Ammonia)

Temp. °F	Pressure PSIG	Temp. °F	Pressure PSIG	Temp. °F	Pressure PSIG	Temp. °F	Pressure PSIG	Temp. °F	Pressure PSIG
−105	27.9*	−55	16.6*	−5	12.2	45	66.3	95	181.1
−104	27.8*	−54	16.2*	−4	12.9	46	67.9	96	184.2
−103	27.7*	−53	15.7*	−3	13.6	47	69.5	97	187.4
−102	27.6*	−52	15.3*	−2	14.3	48	71.1	98	100.6
−101	27.5*	−51	14.8*	−1	15.0	49	72.8	99	193.9
−100	27.4*	−50	14.3*	0	15.7	50	74.5	100	197.2
−99	27.3*	−49	13.8*	1	16.5	51	76.2	101	200.5
−98	27.2*	−48	13.3*	2	17.2	52	78.0	102	203.9
−97	27.1*	−47	12.8*	3	18.0	53	79.7	103	207.3
−96	26.9*	−46	12.2*	4	18.8	54	81.5	104	210.7
−95	26.8*	−45	11.7*	5	19.6	55	83.4	105	214.2
−94	26.7*	−44	11.1*	6	20.4	56	85.2	106	217,8
−93	26.6*	−43	10.6*	7	21.2	57	87.1	107	221.3
−92	26.4*	−42	10.0*	8	22.1	58	89.0	108	225.0
−91	26.3*	−41	9.3*	9	22.9	59	90.9	109	228.6
−30	26.1*	−40	8.7*	10	23.8	60	92.9	110	232.3
−89	26.0*	−39	8.1*	11	24.7	61	94.9	111	236.1
−88	25.8*	−38	7.4*	12	25.6	62	96.9	112	239.8
−87	25.6*	−37	6.8*	13	26.5	63	98.9	113	243.7
−86	25.5*	−36	6.1*	14	27.5	64	101.0	114	247.5
−85	25.3*	−35	5.4*	15	28.4	65	103.1	115	251.5
−84	25.1*	−34	4.7*	16	29.4	66	105.3	116	255.4
−83	24.9*	−33	3.9*	17	30.4	67	107.4	117	259.4
−82	24.7*	−32	3.2*	18	31.4	68	109.6	118	263.5
−81	24.5*	−31	2.4*	19	32.5	69	111.8	119	267.6
−80	24.3*	−30	1.6*	20	33.5	70	114.1	120	271.7
−79	24.1*	−29	0.8*	21	34.6	71	116.4	121	275.9
−78	23.9*	−28	0.0*	22	35.7	72	118.7	122	280.1
−77	23.7*	−27	0.4	23	36.8	73	121.0	123	284.4
−76	23.5*	−26	0.8	24	37.9	74	123.4	124	288.7
−75	23.2*	−25	1.3	25	39.0	75	125.8	125	293.1
−74	23.0*	−24	1.7	26	40.2	76	128.3		
−73	22.7*	−23	2.2	27	41.4	77	130.7		
−72	22.4*	−22	2.6	28	42.6	78	133.2		
−71	22.2*	−21	3.1	29	43.8	79	135.8		
−70	21.9*	−20	3.6	30	45.0	80	138.3		
−79	21.6*	−19	4.1	31	46.3	81	140.9		
−68	21.3*	−18	4.6	32	47.6	82	143.6		
−67	21.0*	−17	5.1	33	48.9	83	146.3		
−66	20.7*	−16	5.6	34	50.2	84	149.0		
−65	20.4*	−15	6.2	35	51.6	85	151.7		
−64	20.0*	−14	6.7	36	52.9	86	154.5		
−63	19.7*	−13	7.3	37	54.3	87	157.3		
−62	19.4*	−12	7.9	38	55.7	88	160.1		
−61	19.0*	−11	8.5	39	57.2	89	163.0		
−60	18.6*	−10	9.0	40	58.6	90	165.9		
−59	18.2*	−9	9.7	41	60.1	91	168.9		
−58	17.8*	−8	10.3	42	61.6	92	171.9		
−57	17.4*	−7	10.9	43	63.1	93	174.9		
−56	17.0*	−6	11.5	44	64.7	94	178.0		

(*) Vacuum = inches Mercury
  PSIG = pounds – per square inch gauge

Wet Bulb Temperature °F	Total Heat BTU/Lb. Air	Wet Bulb Temperature °F	Total Heat BTU/Lb. Air	Wet Bulb Temperature °F	Total Heat BTU/Lb. Air	Wet Bulb Temperature °F	Total Heat BTU/Lb. Air
32	11.758	62	27.85	92	58.78	122	125.98
33	12.169	63	28.57	93	60.25	123	129.35
34	12.585	64	29.31	94	61.77	124	132.8
35	13.008	65	30.06	95	63.32	125	136.4
36	13.438	66	30.83	96	64.92	126	140.1
37	13.874	67	31.62	97	66.55	127	143.9
38	14.319	68	32.42	98	68.23	128	147.8
39	14.771	69	33.25	99	69.96	129	151.8
40	15.230	70	34.09	100	71.73	130	155.9
41	15.697	71	34.95	101	73.55	131	160.3
42	16.172	72	35.83	102	75.42	132	164.7
43	16.657	73	36.74	103	77.34	133	169.3
44	17.149	74	37.66	104	79.31	134	174.0
45	17.650	75	38.61	105	81.34	135	178.9
46	18.161	76	39.57	106	83.42	136	183.9
47	18.680	77	40.57	107	85.56	137	189.0
48	19.211	78	41.58	108	87.76	138	194.4
49	19.751	79	42.62	109	90.03	139	199.9
50	20.301	80	43.69	110	92.34	140	205.7
51	20.862	81	44.78	111	94.72	141	211.6
52	21.436	82	45.90	112	97.18	142	217.7
53	22.020	83	47.04	113	99.71	143	224.1
54	22.615	84	48.22	114	102.31	144	230.6
55	23.22	85	49.43	115	104.98	145	237.4
56	23.84	86	50.66	116	107.73	146	244.4
57	24.48	87	51.93	117	110.55	147	251.7
58	25.12	88	53.23	118	113.46	148	259.3
59	25.78	89	54.56	119	116.46	149	267.1
60	26.46	90	55.93	120	119.55	150	275.3
61	27.15	91	57.33	121	122.72	151	283.6

DEW POINT TEMPERATURES

Temperature	Relative Humidity			
	50	60	70	80
70	50	55	60	64
75	55	61	65	69
80	60	65	70	73
85	64	70	75	78
90	69	74	79	84

If air conditioning system is shut down at night high dew point outside air will wet all inside surfaces that are cooler than that temperature and grow mold and mildew.

Close outside air intake and use back draft dampers on exhaust fans.

Chapter 6

Case Studies, Savings

			Estimated $1000 x Annual Savings	Payback in years
A.	Commercial Buildings			
	1.	Bank and Office 8 stories 100,000 sf	82	1.50
	2.	Office (2) 6, 7 stories 300,000 sf		
		Adjustment	300	-0-
		Retrofit	470	.96
	3.	Office 20 stories, 300,000 sf	200	-0-
	4.	Office 2 stories, 20,000 sf	3	.83
	5.	Resort Area		
		Adjustment	10	-0-
		Phase I	40	.63
		Phase II	500	1.20
	6.	Retail Store	113	.53

		Estimated $1000 x Annual Savings	Payback in years
B.	Hospitals		
	1. 1,500 Bed		
	Adjustments	93	-0-
	Computer Room	6.3	1.52
	Absorption Chiller	200	3.00
	Boiler Plant	200	.50
	2. 1000 Bed		
	Cogeneration	570	7.90
	3. 450 Bed		
	Surgery	5.2	2.00
	Cogeneration	400	4.50
	4. 400 Bed		
	Control Replacement	50	5.20
	5. 360 Bed		
	Controls Replacement	54	6.30
	6. State School		
	Boiler Plant Improvements	42	1.90
	New Boilers	130	2.30
C.	Other Buildings		
	1. High Rise Condominium		
	15 stories, 78 units, 31 years old	850	.35
	2. Condominium		
	5 stories, 117 units, 6 years old		
	Relamping	4.8	1.11
	3. Residence		
	12,000 sf, 13 years old	38	.37
	4. Country Club House		
	22,000 sf, 22 years old	112	1.00
	5. Country Club House		
	25,000 sf, 2 years old	10	-0-

		Estimated $1000 x Annual Savings	Payback in years
6.	Country Club House 38,000 sf, 2 years old	60	.40
7.	Electronic Parts	77	1.60
8.	Electronic Clean Room	72	.50
9.	Computer Data Center Plus 100,000 sf Office Cogeneration	110	.50

Chapter 7

Case Studies, Description

A1. **Bank Building, 8 Stories, 100,000 S.F.**

Lighting

Tests showed lighting levels as high as 160 FC where 80 is acceptable. Louvre ceiling absorbs light energy of 10 watts/s.f. supplying only 20 FC.

Air Conditioning

Central system, not directional zoned, temperature variation 10°. Large refrigeration units required start up at 6AM requiring operator overtime.

Recommendations

Retrofits of lighting, cooling, and controls indicates a savings of $82,000/year.

A2. **Office Building (2)**

This office building of 300,000 s.f. floor area, approximately 18 years old, was tested for operating efficiency on both cooling and heating. The system consisted of five major double duct systems with a total 250,000 CFM capacity with the perimeter of the building being supplied with medium pressure mixing boxes and the interior of the building supplied by mixing dampers. Although the system was operated from a master control panel in another building, with round the clock attendance and the system had reasonable maintenance, it was found through field tests that most of the control devices had deteriorated to where very little control was occurring. The mixing box dampers were breaking at a rate of 5 per day and there were at least 200 thermostats inoperative.

Outside air supply to the building, even though the dampers were tight closed, was still in excess of the Code ventilation requirements and the toilet exhaust fans were operating in excess of actual requirements for this purpose.

The building was air conditioned and heated during the period of cleaning the floors which was performed by six people from 5:30 PM to 12:30 AM five days a week. When asked why the building must be air conditioned during the cleaning operation, the answer was that the wax wouldn't dry. A field test was made with a temperature recorder on a weekend when the daytime outside temperature was 85°F or higher. The temperature rise in the building was only 2°F with the complete air conditioning system shut down for the full weekend period.

The original building was provided with 14 switches on the wall in the corridor for each floor so that the lighting could be zoned during cleaning operations. These switches were not used by the cleaning personnel with the result that the entire floor was lighted. Recommendations were made to re-use these switches and only light that section of the floor which was being cleaned.

The building had 4 elevators, each of which is operated by a motor generator set. All 4 elevators were active at all times. It was recommended that under light load conditions and on nights and weekends, only one elevator be activated and the motor generator sets for the other 3 elevators be shut down.

There is a 27,000 s.f. parking garage in the building supplied with a continuous ventilation system to reduce carbon monoxide. The installation of carbon monoxide detecting equipment cycling the ventilation system in the winter time indicated a one-year payback.

A3. **Office**

This 20 story office building had an energy use of 407,000 BTU/s.f.

System was chilled water produced by steam absorption chillers. The four gas boilers had poor combustion.

Testing indicated negative air pressure in boiler room, created by exhaust fan and absence of outside air intake.

Heating was provided by a heat exchanger in chilled water line. Leaving water thermometer indicated control valve leaking steam to heat exchanger in cooling cycle, Correction: exhaust fan reversed, valve repaired.

A4. **Office**

Tests of air flow indicated unit air supply of 22,000 CFM but total outlet air was 8,000 CFM. Return air in ceiling 4° below room temperature. Sheet metal duct joints cleaned and taped for correction.

A5. **Resort Area**

Air Conditioning

System is primarily 33% outside air with reheat. Total system 3,000,000 CFM, outside air 1,000,000 CFM.

Modification of reheat to zone recirculation indicates savings of $500,000 per year.

A6. **Retail Store**

Large department store divided into four separate stores with common ceiling return but with separate air conditioning units.

Tests showed ceiling air at 65° on 90° day from duct leakage.

Excess outside air supply.

Lights on the parking lot and canopy during daytime. Time clocks not operable.

B1. **1,000 Bed Hospital**

Air Conditioning

Chiller Plant – Combination electric and absorption. Use of steam cost 40% less than electric. Convert system to all absorption.

Computer Center cooled from central system. Separate chiller will let main system shut down in winter.

150 induction units in patient rooms operated from duct pressure thermostats. All systems plugged up with lint.

Maintenance clean-up required only, saving $50,000/year.

Boiler Plant

Poor control and inaccurate instruments on gas boilers.

High feed water heater temperature – flash loss – $28,000/year.

50% more makeup water creating at least 15% more fuel. Direct steam injection used in laundry.

Chemical treatment savings – $15,000/year if makeup water is reduced by one-half.

Incinerator

Adjustment of gas burners and modification of combustion chamber will reduce gas consumption by 20%, permit reducing smoke (cited by EPA), allow burning trash in 8 hours instead of 10 hours.

Heat recovery boiler, cleaned, adjusted, and maintained will produce sufficient steam for 800 tons of cooling.

B2. **1,000 Bed Hospital**

Annual electric cost for hospital building only, $1,600,000. Use of eight 650kw gas engine generators will provide complete standby. Absorption cooling and domestic hot water.

Complete automatic transfer to utility power is included.

Cost of gas engine operation includes maintenance cost. Waste heat produces 12 psi steam.

B3. **450 Bed Hospital**

Air Conditioning

Analysis of surgery, non-occupied, recirculation system indicates a 2 year payback.

Cogeneration

Analysis of use of gas engine generating plant, waste heat absorption cooling, standby equipment, power company automatic transfer indicates a 4-1/2 year payback. Equipment building is included.

B4. **400 Bed Hospital**

Sixty patient rooms conditioned by central system with hot water reheat coils for each room controlled by room thermostat.

Supply air temperature tests showed 53 coil control valves and 46 thermostats defective. 140° heating water overheated reheat coils for 20 minutes after shut off.

Heat wheel on fan intake partially plugged up, with nearly 50% air bypassed and one inch water gauge extra air pressure drop.

Valves and thermostats replaced. Heat wheel removed, system placed on partial recirculation. Air supply balanced.

B5. **360 Bed Hospital**

Recommendations of 19 projects for control and heat recovery including surgery, recovery, kitchen, central supply, dining room and patient rooms.

C1. **High Rise Condominium**

Equipment and piping replacement (no heat)	
Cost	$ 850,000
Fire dampers in corridors	300,000
	$1,150,000
Reuse of 75% of existing units, no extra	
fire dampers required	$ 300,000

System consisted of air handling unit for each apartment, cooling coil, two position control valve, manual summer/winter switch. Hot water heating used same piping as chilled water, but inoperative. Piping insulation defective, piping corroding on outside as well as inside.

Sample unit tests indicated that with some repairs, at least 75% of air conditioning units can be reused.

Existing gas boilers inoperative. Toilet exhaust non-existent below 13th floor. Kitchen exhaust also ineffective.

C2. Condominium, 5 Stories, 117 Units

Replace 60W incandescent lamps with PL7 fluorescent lamps on catwalks, lobbies, and fire stairs.

C3. Residence, Large, Oceanside

A. *Air Conditioning*

1. System consisted of chilled water cooling, hot water heating, both electric. Air handlers had chilled water and separate hot water coils. Control – program type, humidistat over cooling with reheat for moisture removal.

2. Tests indicated chiller producing 4 tons at 24HP, controls defective, electric boiler cycling at 100% capacity.

3. Electric cost before correction: $6,000 per month; after correction: $2,900 per month.

C4. Country Clubhouse

A. *Air Conditioning*

1. Modified duct work.
2. Reused existing air handlers.
3. Replaced 25% of air cooled compressors.
4. Installed remote program thermostats.

5. Electric Bill:
    After Retrofit $3.75/sq.ft./year
    Clubhouse C5
        Before Retrofit $6.36/sq.ft./year
    Clubhouse C6
        Before Retrofit $4.65/sq.ft./year

C5. Country Club Clubhouse

Chilled water system tested low on static pressure. Little water flow to second floor air conditioners. Chilled water temperature 52°.

Corrected water pressure with new 30 psi relief valve, lowered chilled water to 42°. Recommended remote program control, system balanced.

C5. Country Club

Correction recommended: Program control. 400 seat dining room tested at 70°, all lights on at noon with two people eating.

Irrigation system for golf course and grounds indicated 100% fluctuation in monthly electric bills due to poor hydraulic control.

C7. Electronic Parts

Correction of outside air, hot and cold deck controls required.

C8. Electronic Clean Room

Air conditioning system tested at 9,000 CFM, 0° supply. Some ducts partially bypassing supply to return air. Ducts too small.

Correction: Ducts modified, air supply increased to 28,000 CFM at 53°, refrigeration reduced from 228 tons to 76 tons.

C9. Computer Data Center

In a 2,000kw Data Center, a cogenerating plant was not only cheaper than a standby battery pack, but also could operate the center and an adjoining eleven-story office building for heating, cooling, electricity, at half the utility rate.

Chapter 8

Test Data Sheets

TEST DATA – CONDITIONED SPACE

1. Room Temperature
   A   _____
   B   _____
   C   _____
   D   _____
   E   _____
   F   _____

	Static	High	Low
2. Thermostat Setting	_____	_____	_____
3. Air Supply Temperature	_____	_____	_____
4. Air Temperature Above Ceiling			_____
5a. Distance of air outlet to end of its conditioned area			_____

5b. Any auxiliary heat source? What? _____

6a. Ceiling height			_____
6b. Air outlet height			_____
7. Ceiling fan height			_____
Outside Air Temperature D.B.			_____
Outside Air Temperature W.B.			_____

TEST DATA – AIR HANDLING EQUIPMENT

1. Air Supply Temperature   _____

2. Air temperature rise unit to outlet   _____

3. Type of system
    a.   Single Zone   _____
    b.   Multizone   _____
    c.   Reheat   _____
    d.   Mixing Box   _____
    e.   V.A.V.   _____
    f.   Induction   _____

4. R.A. temperature   _____

5. O.A. temperature   _____

6. Mixed air temperature   _____

7. % O.A.   _____

8. Required O.A.   _____

9. Cooling Coil   _____
    a.   Clean   _____
    b.   Return bends cool   _____
    c.   Number of rows   _____

10a. Coil mounting tight   _____

10b. Fan clean   _____

11. Fan belts tight   _____

12. Fan motor

            Amps _____   Rating _____
            Volts _____   Rating _____

13. Filters clean   _____

14. Housing leaks   _____

15. Air static pressure
    a.   Entering   _____
    b.   Leaving   _____

TEST DATA – REFRIGERATION SYSTEM

Condensing Unit

Location of Condenser

Relation to Air Handler   Above_____   Below_____

Electric   Amps   Volts

Compressor No. 1   _____   _____
Nameplate   _____   _____

Compressor No. 2   _____   _____
Nameplate   _____   _____

Number of Condenser Fans   _____
Number of Fans Running   _____
Sight Glass Clear?   _____
Short Cycle   _____
Rated Capacity (tons)   _____

Water Cooled Units

(2a) Water temperature entering   _____

Water temperature leaving   _____

Water pump pressure   _____

Suction   _____
Discharge   _____

Air Cooled Units

(1a) Air temperature entering   _____

Air temperature leaving   _____

Number of fans   _____

Number of fans running   _____

TEST DATA – CHILLED WATER SYSTEMS

1. Vertical height of highest coil above pressure
    measurement location. Ft. 2.4 ft. – 1 psi   _____

2. Pump Suction Pressure   _____

3. Pump Discharge Pressure   _____

4. Control Valves on Cooling Coils
    a.   Two Way   _____
    b.   Three Way Bypass   _____

5. Chilled Water Pump
    Amps   Actual _____   Rated _____
    Volts   Actual _____   Rated _____
    Size Suction   _____
    Discharge   _____
    Motor Speed RPM   _____

6. Chilled Water Pressure Farthest
    Point From Pumps   Pressure _____
                                  Temp.   _____

7. Pump Bypass Type _____

TEST DATA – DUCT SYSTEM

1. Static Pressure   Entering   Leaving

     a.   Turning vanes   _____   _____

     b.   Fire dampers   _____   _____

     c.   Manual damper   _____   _____

     d.   Mixing Box   _____   _____

     e.   Behind outlets   _____   _____

     f.   Reheat Coils   _____   _____

     g.   Dampers Condition _____

2. Motor Operation _____

3. Air Flow Balance   Spec.   Actual

     a.   First outlet   _____   _____

     b.   Last outlet   _____   _____

4. Duct Construction

     a.   Sheet Metal   _____

     b.   Fiberglass   _____

     c.   Condition _____

     d.   Outlet Tap   Sheet metal_____   Fiberglass_____

         1.   Is Branch Flexible Ducts   _____

             Metal   _____

             Fiberglass   _____

TEST DATA – CONTROLS

1. Electric
   a. ON-OFF   _____
   b. Modulating   _____
   c. MFGR   _____
   d. Thermostat Accuracy   _____

2. Pneumatic
   a. ON-OFF   _____
   b. Gradual Acting   _____
   c. MFGR   _____
   d. Air Line Pressure   _____
   e. Compressor After cooler   _____
   f. Thermostat Condition   _____

3. Electronic
   a. ON-OFF   _____
   b. Modulating   _____
   c. MFGR   _____
   d. Thermostat   _____
        In Room   _____
        Remote Sensor   _____

4. Refrigeration Compressor
     Control

5. Heating
     Electric, No. of Steps   _____
     Hot Water Modulating?   _____
     Steam   _____

TEST DATA – COOLING TOWER

		Dry Bulb	Wet Bulb
1.	Air Temperatures		
	a. Outside	_____	_____
2.	Tower Fill Condition	_____	
3.	Drain Pan Water Level	_____	
4.	Fans		
	a. Number	_____	
	b. Size	_____	
	c. Speed	_____	
	d. Control	_____	
5.	Overflow Quantity	_____	
6.	Plume	_____	
7.	MFGR	_____	
8.	Construction		
	a. Housing	_____	
	b. Fill	_____	
9.	Proximity to Obstruction		
	a. Distance	_____	
	b. Type	_____	
	c. Size	_____	
10.	Location	_____	

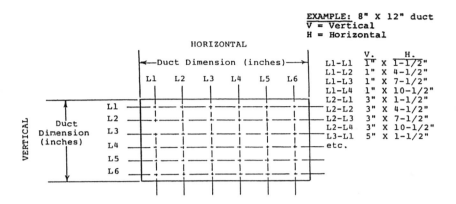

To use table: Measure inside duct size horizontal and vertical. Chart shows test points as a combination of both dimensions. (See example.)

Duct Dim. Inches	Distances From Inner Wall – Inches							
	L1	L2	L3	L4	L5	L6	L7	L8
8	1	3	5	7	---	–	–	–
10	1-1/4	3-3/4	5-1/4	8-3/4	–	–	–	–
12	1-1/2	4-1/2	7-1/2	10-1/2	–	–	–	–
24	2	6	10	14	18	22	–	–
36	3	9	15	21	27	33	–	–
48	4	12	20	28	36	44	–	–
60	5	15	25	35	45	55	–	–
72	4-1/2	13-/12	22-1/2	31-1/2	40-1/2	49-1/2	58-1/2	67-1/2
84	5-1/4	15-3/4	26-1/4	36-3/4	47-1/4	57-3/4	68-1/4	78-3/4
96	6	18	30	42	54	66	78	90
108	6-3/4	20-1/4	33-3/4	47-1/4	60-3/4	74-1/4	87-3/4	101-1/4
120	7-1/2	22-1/2	37-1/2	52-1/2	67-1/2	82-1/2	97-1/2	112-1/2
132	8-1/4	24-3/4	41-1/4	57-3/4	74-1/4	90-3/4	107-1/4	123-3/4
144	9	27	45	63	81	99	117	135
156	9-3/4	29-1/4	48-3/4	68-1/4	87-3/4	107-1/4	126-3/4	146-1/4

Figure 8-1   Location of Air Velocity Readings in Square or Rectangular Ducts

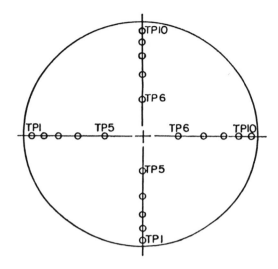

Distance along diameter from inner wall to test point (TP):

Test Point No.	Dimension (inches)	Test Point No.	Dimension (inches)
1	1/4	6	6-1/2
2	7/8	7	7-3/4
3	1-1/2	8	8-1/2
4	2-1/4	9	9-1/8
5	3-1/2	10	9-3/4

Figure 8-2 Test Point Locations for Circular Points
Duct diameter 10 inches

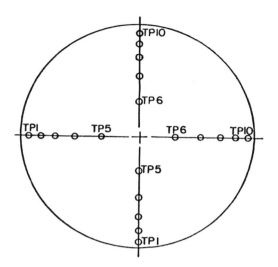

Distance along diameter from inner wall to test point (TP):

Test Point No.	Dimension (inches)	Test Point No.	Dimension (inches)
1	3/8	6	7-7/8
2	1	7	9-1/4
3	1-3/4	8	10-1/4
4	2-3/4	9	11
5	4-1/8	10	11-5/8

Figure 8-3   Test Point Locations for Circular Points
Duct diameter 12 inches

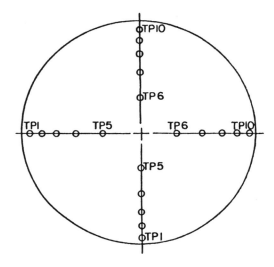

Distance along diameter from inner wall to test point (TP):

Test Point No.	Dimension (inches)	Test Point No.	Dimension (inches)
1	1/2	6	11-7/8
2	1-1/2	7	14
3	2-5/8	8	15-3/8
4	4	9	16-1/2
5	6-1/8	10	17-1/2

Figure 8-4 Test Point Locations for Circular Points
Duct diameter 18 inches

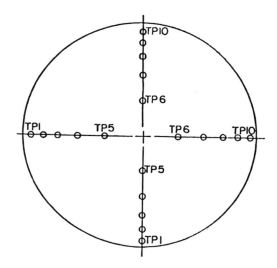

Distance along diameter from inner wall to test point (TP):

Test Point No.	Dimension (inches)	Test Point No.	Dimension (inches)
1	5/8	6	15-3/4
2	2	7	18-1/2
3	3-1/2	8	20-1/2
4	5-1/2	9	22
5	8-1/4	10	22-3/8

Figure 8-5 Test Point Locations for Circular Points
Duct diameter 24 inches

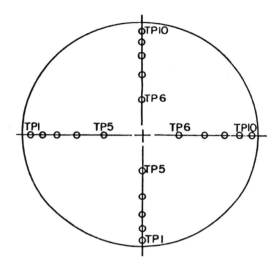

Distance along diameter from inner wall to test point (TP):

Test Point No.	Dimension (inches)	Test Point No.	Dimension (inches)
1	1	6	23-3/4
2	3	7	27-7/8
3	5-1/4	8	30-3/4
4	8-1/8	9	33
5	12-1/4	10	35

Figure 8-6   Test Point Locations for Circular Points
Duct diameter 36 inches

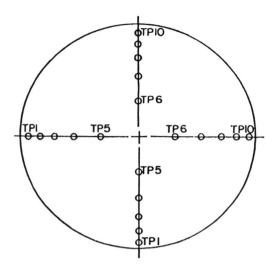

Distance along diameter from inner wall to test point (TP):

Test Point No.	Dimension (inches)	Test Point No.	Dimension (inches)
1	1-1/4	6	31-1/2
2	4	7	37-1/8
3	7	8	41
4	10-7/8	9	44
5	16-1/2	10	46-3/4

Figure 8-7 Test Point Locations for Circular Points
Duct diameter 48 inches

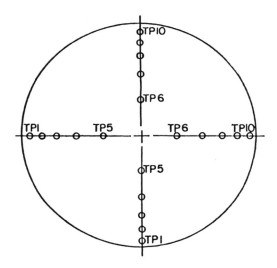

Distance along diameter from inner wall to test point (TP):

Test Point No.	Dimension (inches)	Test Point No.	Dimension (inches)
1	1-1/2	6	39-1/2
2	4-7/8	7	46-1/2
3	8-3/4	8	51-1/4
4	13-1/2	9	55-1/8
5	20-1/2	10	58-1/2

Figure 8-8 Test Point Locations for Circular Points
Duct diameter 60 inches

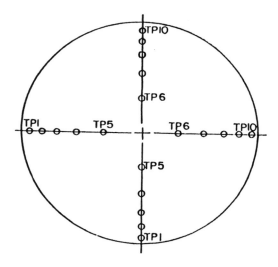

Distance along diameter from inner wall to test point (TP):

Test Point No.	Dimension (inches)	Test Point No.	Dimension (inches)
1	1-7/8	6	47-3/8
2	5-7/8	7	55-3/4
3	10-1/2	8	61-1/2
4	16-1/4	9	66-1/8
5	24-5/8	10	70-1/8

Figure 8-9   Test Point Locations for Circular Points
Duct diameter 72 inches

AIR OUTLET MEASUREMENTS

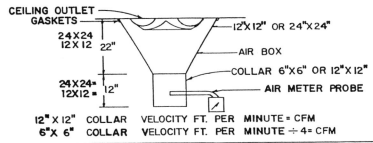

CEILING OUTLET
GASKETS
24X24
12X12 22"

12"X12" OR 24"X24"

AIR BOX

COLLAR 6"X6" OR 12"X12"

24X24=
12X12 = 12"

AIR METER PROBE

12" X 12" COLLAR VELOCITY FT. PER MINUTE = CFM
6"X 6" COLLAR VELOCITY FT. PER MINUTE ÷ 4= CFM

SIDE WALL SUPPLY GRILLE, FLOOR GRILLE
(SIMILAR)

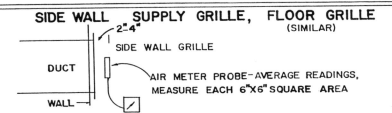

2"-4"

SIDE WALL GRILLE

DUCT

AIR METER PROBE-AVERAGE READINGS,
MEASURE EACH 6"X6" SQUARE AREA

WALL

CFM= AVERAGE VELOCITY FT/MINUTE X SQ. FT. OF GRILLE INSIDE FRAME

RETURN AIR GRILLE, OUTDOOR AIR INTAKE

24"

BOX SAME DIMENSIONS AS GRILLE
OPENING OR LARGER

DUCT

AIR METER PROBE- AVERAGE READINGS,
MEASURE EACH 6"X6" AREA

WALL

CFM= AVERAGE VELOCITY FT/MINUTE X SQUARE FEET AREA OF BOX FACE

Figure 8-10 Air Outlet Measurements

Chapter 9

General Field Information

Building: _____

    Location _____

    Configuration _____

    Construction: _____

            Wall _____

            Roof _____

         Windows _____

           Doors _____

Condition _____

_____

_____

Type of Heating System _____

Type of Air Conditioning System

Occupancy Periods _____ Hrs./Day _____ Days/Wk. _____ Wks./Yr.

Average Number of Occupants per Day_____

Building Location:

    A. Separate and isolated

    B. Part of group of multi-story buildings
        (Amount of exposure)

    C. Basement space

Sq. Ft. of Floor Area _____ Occupied _____ Storage

Number of Floors _____ Occupied _____ Storage

Gross Sq. Ft. of Floor Space_____

ENERGY CONSUMPTION FOR PREVIOUS YEAR

19____	Electricity		Natural Gas		Fuel Oil Type		Coal – Heating Value (BTU/#)		Propane	
	Use (Kwh)	Cost	Use (MCF)	Cost	Use (Gal)	Cost	Use (Tons)	Cost	Use (Gal)	Cost
January		$		$		$		$		$
February		$		$		$		$		$
March		$		$		$		$		$
April		$		$		$		$		$
May		$		$		$		$		$
June		$		$		$		$		$
July		$		$		$		$		$
August		$		$		$		$		$
September		$		$		$		$		$
October		$		$		$		$		$
November		$		$		$		$		$
December		$		$		$		$		$
Total		$		$		$		$		$
Average Cost	$/KWH	$	$/MCF	$	$/Gal.	$	$/Ton	$	$/Gal	$

GENERAL FIELD INFORMATION

FI-1   Type of Operation   _____

FI-2   Building Height   _____

FI-3   Square Feet Floor Area   _____

FI-4   Roof Area Square Feet   _____

FI-5   Wall Area Square Feet   _____

FI-6   Window Area Square Feet   _____

FI-7   Toilet Rooms Area Square Feet   _____

FI-8   Building Location:

       A.   Separated or isolated   _____

       B.   Part of a group building   _____

       C.   Part of a multi-story building   _____

       D.   Below ground level   _____

FI-9   Hours of Operation – Daily, Weekly, Yearly _____

FI-10 Number of people occupying space   _____

FI-11 Specific Legal or Health Requirements

       A.   Ventilation   _____

       B.   Temperature   _____

       C.   Smoking   _____

       D.   Air Contaminants   _____

       E.   Special Building Codes   _____

FI-12 Visual Tasks (light level requirements)   _____

COMMERCIAL

	KW Ratings	or BTU/hr.	Misc.
FI-13 Equipment Requirements			
A. Domestic Hot Water	_____	_____	_____
B. Refrigeration	_____	_____	_____
C. Computers	_____	_____	_____
D. Cooking and Baking	_____	_____	_____
E. Office Machines	_____	_____	_____
F. Cash Registers	_____	_____	_____
G. Food Processing (Slicers, grinders)	_____	_____	_____
H. Elevators	_____	_____	_____
I. Escalators	_____	_____	_____
J. Water Pumps	_____	_____	_____
K. Sewage Pumps	_____	_____	_____
L. Incinerators	_____	_____	_____
M. Lighting – Outside			
(1) Flood	_____	_____	_____
(2) Sign	_____	_____	_____
(3) Parking Lot	_____	_____	_____
(4) Security	_____	_____	_____
N. Heating System			
(1) Type	_____	_____	_____
(2) Fuel	_____	_____	_____
(3) Heating Capacity	_____	_____	_____
(4) Motors	_____	_____	_____
O. Cooling System			
(1) Type	_____	_____	_____
(2) Fuel	_____	_____	_____
(3) Cooling Capacity	_____	_____	_____
(4) Motors	_____	_____	_____

FI-14 Employee Requirements

A. Psychological   _____

B. Union Regulations   _____

FIELD SURVEY

I. Office

    A. Building Heat Transmission

        1. Roof

            a. Construction _____ ___ ___ ___
            b. Insulation _____ ___ ___ ___
            c. Roof color_____ ___ ___ ___
            d. Roof spray_____ ___ ___ ___
            e. Flooded roofs _____ ___ ___ ___

        2. Walls

            a. Construction _____ ___ ___ ___
            b. Insulation _____ ___ ___ ___
            c. Color_____ ___ ___ ___
            d. Solar exposure _____ ___ ___ ___

        3. Glass

            a. Windows_____ ___ ___ ___
            b. Solar exposure _____ ___ ___ ___
            c. Skylights_____ ___ ___ ___
            d. Canopies _____ ___ ___ ___
            e. Reflecting surfaces _____ ___ ___ ___
            f. Plastics _____ ___ ___ ___
            g. Window covering _____ ___ ___ ___

        4. Air leakage

            a. Infiltration_____ ___ ___ ___
            b. Building exhaust_____ ___ ___ ___
            c. Window leakage _____ ___ ___ ___
            d. Wall leakage _____ ___ ___ ___
            e. Door leakage_____ ___ ___ ___
            f. Elevator shafts _____ ___ ___ ___
            g. Stairways_____ ___ ___ ___
            h. Vestibules _____ ___ ___ ___
            i. Wind velocity _____ ___ ___ ___

I. Office (continued)

    B. Internal Heat Gain

        1. Lights

            a. Light level_____ ___ ___ ___

            b. Reuse lighting heat in winter_____ ___ ___ ___

            c. Removing lighting heat in summer __ ___ ___ ___

        2. People

            a. Area concentration _____ ___ ___ ___

            b. Active _____ ___ ___ ___

            c. Inactive _____ ___ ___ ___

        3. Equipment

            a. Office machines _____ ___ ___ ___

            b. Computers_____ ___ ___ ___

            c. Vending machines_____ ___ ___ ___

        4. Environment

            a. Interior color_____ ___ ___ ___

            b. Space arrangement _____ ___ ___ ___

            c. Proximity of seated persons to
               outside walls_____ ___ ___ ___

        5. Power

            a. Self-contained drinking fountains ___ ___ ___ ___

            b. Electric power_____ ___ ___ ___

    C. External Energy Use

        1. Water booster pumps_____ ___ ___ ___

        2. Elevators _____ ___ ___ ___

    D. Heating and Air Conditioning

II. Education Facilities

   A. Building Heat Transmission _____ ___ ___ ___

   B. Classroom

      1. Ventilation _____ ___ ___ ___

      2. Heating _____ ___ ___ ___

      3. Air conditioning _____ ___ ___ ___

      4. Lighting _____ ___ ___ ___

   C. Physical Education

   D. Laboratories

      1. Exhaust – fumes, odor _____ ___ ___ ___

      2. Hot water faucet control _____ ___ ___ ___

      3. Fume hoods _____ ___ ___ ___

      4. Special requirements for temperature
control _____ ___ ___ ___

      5. Electric demand controls _____ ___ ___ ___

   E. Home Economics

      1. Kitchen exhaust _____ ___ ___ ___

      2. Washer and dryer exhausts _____ ___ ___ ___

      3. Sink hot water faucet control _____ ___ ___ ___

   F. Auditoriums

      1. Ventilation _____ ___ ___ ___

      2. Heating and air conditioning _____ ___ ___ ___

   G. Offices _____ ___ ___ ___

   H. Shops _____ ___ ___ ___

III. Dining Facilities

   A. Building Heat Transmission _____ ___ ___ ___

   B. Internal Heat Gain _____ ___ ___ ___

      1. People _____ ___ ___ ___

      2. Food serving equipment _____ ___ ___ ___

         a. Open steam tables _____ ___ ___ ___

         b. Coffee makers _____ ___ ___ ___

         c. Vending machines

III. Dining Facilities (continued)

   B. Internal Heat Gain (continued)

      3. Ventilation

         a. Control _____   ___   ___   ___

         b. Odor removal equipment _____   ___   ___   ___

         c. Exhaust through kitchen_____   ___   ___   ___

      4. Lighting

         a. Light level_____   ___   ___   ___

IV. Kitchens

   A. Building Heat Transmission_____   ___   ___   ___

   B. Internal Heat Gain

      1. Ranges

         a. Fuel type combustion efficiency ____   ___   ___   ___

         b. Electric element efficiency_____   ___   ___   ___

         c. Oven insulation_____   ___   ___   ___

      2. Bake ovens

         a. Time cycle _____   ___   ___   ___

         b. Loading _____   ___   ___   ___

         c. Fuel type combustion efficiency ____   ___   ___   ___

         d. Venting of flue gases _____   ___   ___   ___

      3. Steamers

         a. Steam kettles_____   ___   ___   ___

         b. Shelf type _____   ___   ___   ___

         c. Steam traps _____   ___   ___   ___

         d. Fuel fired self-contained steam
            generators _____   ___   ___   ___

      4. Refrigerators

         a. Freezers_____   ___   ___   ___

         b. Coolers _____   ___   ___   ___

         c. Combination _____   ___   ___   ___

         d. Refrigeration equipment_____   ___   ___   ___

IV. Kitchens (continued)

    B. Internal Heat Gain (continued)

        5. Ventilation

            a. Range hoods _____ ___ ___ ___

            b. Dishwasher hoods_____ ___ ___ ___

            c. General ventilation _____ ___ ___ ___

            d. Heat recovery _____ ___ ___ ___

            e. Contaminant removal _____ ___ ___ ___

            f. Odor removal _____ ___ ___ ___

            g. Ventilating equipment_____ ___ ___ ___

        6. Dishwashers

            a. Location _____ ___ ___ ___

            b. Exhaust connection_____ ___ ___ ___

            c. Booster water heater_____ ___ ___ ___

            d. Detergent use _____ ___ ___ ___

            e. Operation cycle_____ ___ ___ ___

        7. Sinks

            a. Control of hot water_____ ___ ___ ___

            b. Operation cycle_____ ___ ___ ___

            c. Faucet type _____ ___ ___ ___

        8. Lighting

            a. Light level_____ ___ ___ ___

V. Recreation Facilities

    A. Gymnasiums

        1. Lighting _____ ___ ___ ___

        2. Ventilation

            a. Play area _____ ___ ___ ___

            b. Locker rooms _____ ___ ___ ___

        3. Sanitary facilities_____ ___ ___ ___

        4. Drying rooms_____ ___ ___ ___

        5. Physical therapy equipment _____ ___ ___ ___

V.   Recreation Facilities (continued)

    B.   Swimming Pools

       1.   Water heating _____   ___   ___   ___

       2.   Water filtering_____   ___   ___   ___

       3.   Ventilation_____   ___   ___   ___

       4.   Humidity control_____   ___   ___   ___

    C.   Auditoriums

       1.   Heating and air conditioning _____   ___   ___   ___

       2.   Lighting_____   ___   ___   ___

       3.   Stage lighting _____   ___   ___   ___

VI.   Living and Sleeping Areas

    A.   Building Heat Transmission _____   ___   ___   ___

    B.   Heating and Air Conditioning _____   ___   ___   ___

    C.   Furniture Arrangement _____   ___   ___   ___

    D.   Lighting_____   ___   ___   ___

    E.   Boiler Facilities_____   ___   ___   ___

    F.   Laundries_____   ___   ___   ___

VII.   Laundries

    A.   Washers

       1.   Water flow control _____   ___   ___   ___

       2.   Detergent use _____   ___   ___   ___

       3.   Waste water heat recovery _____   ___   ___   ___

       4.   Time cycle _____   ___   ___   ___

    B.   Dryers

       1.   Temperature control _____   ___   ___   ___

       2.   Time cycle _____   ___   ___   ___

       3.   Vent waste heat recovery _____   ___   ___   ___

    C.   Ironers

       1.   Temperature control _____   ___   ___   ___

       2.   Time cycle _____   ___   ___   ___

       3.   Vent hoods _____   ___   ___   ___

       4.   Vent heat recovery

VII. Laundries (continued)

    D. Ventilation

        1. Heat recovery_____ ___ ___ ___

        2. Location of air supply _____ ___ ___ ___

    E. Lighting

VIII. Workshops

    A. Building Heat Transmission_____ ___ ___ ___

    B. Internal Heat Gain

        1. Lights

            a. Light level _____ ___ ___ ___

            b. Reuse lighting heat in winter _____ ___ ___ ___

            c. Removal of lighting heat
               in summer _____ ___ ___ ___

        2. People

            a. Area concentration _____ ___ ___ ___

            b. Active_____ ___ ___ ___

        3. Equipment

            a. Motors _____ ___ ___ ___

            b. Furnaces_____ ___ ___ ___

            c. Electric appliances_____ ___ ___ ___

        4. Ventilation

            a. For suspended air contamination ___ ___ ___ ___

            b. For comfort _____ ___ ___ ___

            c. For concentrated air contamination _ ___ ___ ___

            d. Toilet facilities_____ ___ ___ ___

    C. Air Conditioning

        1. Type system

            a. Single zone _____ ___ ___ ___

            b. Multizone_____ ___ ___ ___

            c. Zone reheat _____ ___ ___ ___

            d. Medium pressure mixing box_____ ___ ___ ___

VIII. Workshops (continued)

    C. Air Conditioning (continued)

        1. Type system (continued)

           e. High pressure induction _____ ___ ___ ___

           f. Variable volume_____ ___ ___ ___

        2. Filters

           a. Type _____ ___ ___ ___

           b. Cleanliness_____ ___ ___ ___

           c. Air pressure drop_____ ___ ___ ___

        3. Equipment

           a. Motors _____ ___ ___ ___

           b. Furnaces_____ ___ ___ ___

           c. Electric appliances_____ ___ ___ ___

           d. Dust collectors_____ ___ ___ ___

           e. Battery charging _____ ___ ___ ___

           f. Welding _____ ___ ___ ___

           g. Machinery color _____ ___ ___ ___

           h. Demand control
              (see Section XII) _____ ___ ___ ___

        4. Ventilation

           a. For suspended air contamination ___ ___ ___ ___

           b. For comfort _____ ___ ___ ___

           c. For concentrated air
              contamination _____ ___ ___ ___

           d. Toilet facilities_____ ___ ___ ___

IX. Heating and Air Conditioning

    A. Furnaces and Indirect Fuel Fired Heaters

        1. Fuel combustion efficiency _____ ___ ___ ___

        2. Air flow pattern _____ ___ ___ ___

        3. Temperature of flue gases _____ ___ ___ ___

IX. Heating and Air Conditioning (continued)

   B. Boilers

      1. Combustion efficiency_____ ___ ___ ___

      2. Steam

         a. Pressure_____ ___ ___ ___

         b. Condensate return_____ ___ ___ ___

         c. Heat transfer (mineral deposits)_____ ___ ___ ___

         d. Draft _____ ___ ___ ___

         e. Insulation _____ ___ ___ ___

      3. Hot water

         a. Temperature _____ ___ ___ ___

         b. Pressure_____ ___ ___ ___

         c. Heat transfer_____ ___ ___ ___

         d. Draft _____ ___ ___ ___

         e. Insulation _____ ___ ___ ___

         f. Water flow _____ ___ ___ ___

   C. Systems

      1. Steam heat

         a. Controls_____ ___ ___ ___

         b. Traps _____ ___ ___ ___

         c. Return water _____ ___ ___ ___

         d. Piping pitch_____ ___ ___ ___

         e. Condensate pumps _____ ___ ___ ___

         f. Pipe insulation _____ ___ ___ ___

      2. Coils

         a. Type _____ ___ ___ ___

         b. Cleanliness _____ ___ ___ ___

         c. Air distribution_____ ___ ___ ___

IX. Heating and Air Conditioning (continued)

   C. Systems (continued)

      3. Refrigeration

         a. Type machine _____ __ __ __

            (1) Reciprocating _____ __ __ __

            (2) Centrifugal _____ __ __ __

            (3) Absorption _____ __ __ __

         b. Capacity control _____ __ __ __

         c. Refrigerant temperature _____ __ __ __

         d. Condensing equipment _____ __ __ __

            (1) Air cooled _____ __ __ __

            (2) Cooling towers _____ __ __ __

      4. Controls

         a. Night shut down _____ __ __ __

         b. Economizer _____ __ __ __

         c. Air temperature _____ __ __ __

         d. Zoning _____ __ __ __

         e. Specific occupancy
recommendations _____ __ __ __

            (1) Offices _____ __ __ __

            (2) Education facilities _____ __ __ __

            (3) Dining facilities _____ __ __ __

            (4) Kitchens _____ __ __ __

            (5) Recreation facilities _____ __ __ __

            (6) Living and sleeping _____ __ __ __

            (7) Laundries _____ __ __ __

            (8) Workshops _____ __ __ __

            (9) Toilets and showers _____ __ __ __

         f. Computer monitoring
and control _____ __ __ __

X. Toilets and Showers

    A. Sink Faucets _____ ___ ___ ___

    B. Shower Flow Control_____ ___ ___ ___

    C. Hot Water Temperature _____ ___ ___ ___

    D. Hot Water Heaters

        1. Direct fired _____ ___ ___ ___

        2. Indirect heaters _____ ___ ___ ___

    E. Ventilation_____ ___ ___ ___

    F. Temperature _____ ___ ___ ___

    G. Special Fixtures

        1. Wash fountains _____ ___ ___ ___

        2. Gang showers _____ ___ ___ ___

Chapter 10

Heating

STEAM SYSTEMS

a. Pressure

Steam temperature varies with pressure. Steam can also be super-heated above the saturation temperature. Saturation temperatures at gauge pressures at sea level are: (P.S.I. = pound per square inch)

Pressure PSI Gauge	Temperature °F
0	212
5	227
10	240
20	259
30	275
40	287
50	298
100	338
200	387

If the system can be Operated at a lower pressure, the steam pipe will be cooler and there will be less pipe heat loss even if insulated.

b. Pipe Insulation

Piping containing steam, condensate or hot water will lose heat to the surrounding air or building surface. Piping should have some insulation. Determining whether there is enough requires an eco-nomic analysis. Adding insulation over one to two inch thick may not pay off. For example: 2" pipe at 250° temperature, with 1" insulation, loses 34.5 BTU/hr. per lineal foot; 2" insulation – 22.5 BTU; 3" insulation – 18.5 BTU.

It is more important that the insulation be dry, sealed, and fittings and valves covered.

c. Traps

Steam traps prevent steam entering the condensate line, reducing the capacity for water removal and increasing the pipe temperature. If condensate line is not insulated, much heat will be lost. If condensate line is insulated and all of the equipment is working, use a thermometer in the condensate tank. When this temperature reaches 200°F, start looking for bad traps. Below this temperature, small trap leaks may not hurt system.

d. Condensate temperature

If return water collects in a vented tank, the water should not exceed 200°F. Above this temperature, water begins to flash into steam and is wasted out the vent. If steam pressure is higher than 10 pounds, condensate may be higher than 200°F. Use a coil in the tank and preheat domestic water to lower the condensate temperature.

e. Condensate and vacuum pumps

On small systems, condensate pumps should not run more than 30% of the time. Duplex units, with an automatic alternator, assures minimum failure. Vacuum pumps used on systems requiring condensate lift should be inspected often. Replace or calibrate vacuum gauges regularly. Many vacuum pumps run continuously, due to air leaks into piping or control problems. Sucking air into system also causes corrosion.

f. Controls

(1) The best control of steam systems is to keep a constant pressure and let individual heating devices be controlled by automatic valves. This will let the burner run only when steam is being used.

(2) If the system has no controls on heating devices, the entire system can be time cycled or modulated. On-off time cycling is best, since full steam pressure is available to the end of the supply piping. When modulated, which meters the steam flow, heaters near the main feed will still be too hot and the end of

the line too cold. Unless balancing valves are installed on each heater, add adjusted, the system will fluctuate from one end to the other. Overheating the entering end to heat the far end wastes fuel.

Hot Water Systems

a. <u>Temperature</u>

Normal hot water heating systems operate between 200° and 240° at the boiler, with water returning 20° lower. High temperature hot water systems can run up to 370°, but must be operated at 300 pounds pressure to keep from flashing into steam and vapor locking pumps. High temperature hot water (HTHW) circulates less water, since temperature drops as high as 140° can be used instead of 20°. If your system is operating at less than 10° drop, between boiler outlet and return, consider reducing pump flow and horsepower.

b. <u>Circulating Pumps</u>

Good control on this system makes continuous water flow mandatory, especially if heater control valves are three-way. Under light load conditions, such as at night, in spring and autumn, water flow could be reduced, provided the outside air stays above freezing. Do this by by-passing pump outlet to inlet with a control valve, or by variable speed. Starting and stopping multiple pumps produces the best savings.

c. <u>Pipe Insulation</u>

As with steam systems, pipe heat loss is great from poorly insulated piping. In this case, the return line must be insulated, since it is almost the same temperature as the supply. Pump bodies should also be insulated. Heat loss from pipes is the same as steam at equivalent temperatures.

d. <u>Water Pressure</u>

On buildings three stories or less, city water pressure through a reducing valve usually will put the water to the top floor. Most feeder valves come from the factory set at 12 pounds. Best results, with minimum venting, can be obtained with 20 to 22 pounds pressure, even on one story buildings. Remember that the water pressure must be high enough so that at least 10 pounds is at the highest point in the system, or there is an open expansion tank at the top.

e. Controls

Continuous flow, with the water temperature being changed either by outside or inside air temperature, provides more even heat. This is particularly true if radiation is long fin tube sections. If water flow is interrupted by starting and stopping pumps, or opening and closing valves, periods of cold air feeling will result. Also, if heating element is long and goes through several rooms, the first-fed section will always be hotter than the end section. This type control wastes heat because of uneven distribution.

Heat Exchangers

a. Domestic Water

Steam and hot water is used to heat domestic water, either in a direct heat exchanger, or in a coil submersed within the water in a storage tank. It is essential that this tubing be kept clean from minerals or impurities and requires tube cleaning. In the case of domestic water heaters, with the return tube type heat exchanger, the use of steam within the tubes and the water surrounding the tubes, has a tendency to crack the minerals off the outside of the tube into the bottom of the tank, thus staying cleaner longer.

Piping

Piping for steam and hot water systems is designed and sized for the pressure drop in the system and for the proper flow to various heating devices. This piping should be pitched to drain, especially with condensate, as the water condensed should flow reasonably free to the condensate receiver. Intermediate receivers and pumps should be used to return the water to the boiler plant if the piping cannot be pitched all the way back to the plant. Pipe expansion should be provided for by means of loops or expansion joints to prevent the pipe from breaking at certain points due to expansion when the pipe is heated. Vacuum systems can also lift water into the condensate receiver.

Radiators

Radiators are usually heavy cast iron and give off heat by radiation. They are slow to heat and cool and work best when out in the open. Recessing makes it work like a connector and reduces its capacity. Dark color and rough surface radiates best. Most cast iron radiators are old. Feel the bottom of the sections. If they are cooler, they are probably plugged up with mud or scale.

Radiant panels should have limited surface temperatures, not exceeding 125° at ceiling, 90° on wall, 85° in floor. If heat piping is buried in masonry construction, time lag of surface temperature change can be as much as 8 hours. This heat is best as an auxiliary rather than the prime source.

Radiant pipes, either steam, hot water, or direct fired, should be placed so that no flammable material can come in contact. Paper and wood start to char at 155°F. High surface temperatures concentrate radiation so reflectors may be necessary.

Convectors

Convectors work by causing air to flow vertically over the heating element as warm air rises. Element must be kept clean to prevent resistance to this air flow. The higher the distance from the element to the cabinet top, the greater the capacity.

Unit Heaters

a. Unit heaters force warm air toward the floor and at a distance. The air velocity and type of discharge connection can determine the pattern. Delivery air temperature should not be less than 90° to reduce drafts.

b. Fan coil cabinet units can be fan cycled. Use a surface thermostat on the leaving pipe of the coil so that the fan won't start until the coil is hot. This will eliminate cold drafts on start up.

Central Air Systems

Steam or hot water coils should be kept clean inside tubes and outside, between fins. Leaving air temperature should be controlled at a maximum of 130° and a minimum of 50°. Continuous air supply with modulating heat flow is best. Fan cycling should be done only when coil is hot.

Steam Traps

Steam traps should be taken apart at least once a year and examined for dirt. Remove strainers and clean. If the steam traps do not hold, the temperature of the pipe entering and leaving should be the same. If the trap is working, the leaving temperature should be slightly colder than the entering temperature. If the steam trap does not hold, remove tie thermostatic element, place it in cold water and hot water to see if it moves. If the bellows do not change shape, replace the element.

Air Vents

All hot water heating systems contain air which is driven out of the water when it is heated up to temperature. The air must be vented from all high points, either automatically or with a manual valve. Unless air is vented, coils or radiators may be only partially full and not operate at full capacity. Air also affects steam systems in same manner.

Control Valves

Check control valves for tight shut off. Thermostats should position valves according to the setting. Measure temperature of heater with valve open and closed and compare with entering steam or hot water temperature.

If the heating element is warm for any length, the valve is leaking. Solenoid valves should be checked to determine if they actually close and open. If they close – do they leak? Check similar to shut-off valve. Balancing valves should be checked for excessive heating from certain heating elements such as radiators or convectors and lower heating from other devices. Adjust these by use of a thermometer at the heat emission side of the radiator, convectors or heating unit.

100% Makeup Air Units

To prevent coil freeze up, steam or hot water must not be allowed to get too cold at any one spot. Air distribution should be adjusted to be even over the face of the coil. Safety shut down thermostats, should have at least 25 foot long tubes located partially near the bottom. Outside air dampers must close tight. Use partial recirculated air if possible.

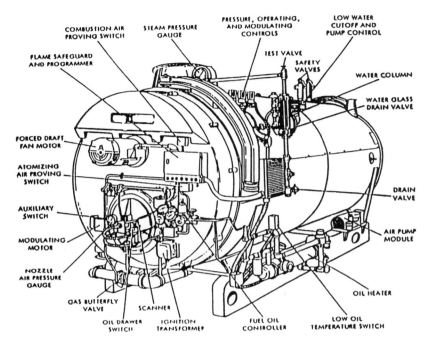

Figure 10-1 (courtesy of Cleaver Brooks)

BOILERS

Energy Management - Heating Systems

1. Boilers

    Boilers may be cast iron, steel fire tube, steel water tube, electric tank type; burners may be gravity, power blower, rotary spray, atomizing, electric resistance, electric conduction probe. The following diagrams indicate energy waste problems to examine, test and correct. Numbers indicate problems and solutions.

2. Energy Waste Problems
    a. Soot in tubes and gas passages
    b. Mud in water at bottom
    c. Hand hole and man hole leaks
    d. Wet insulation

e. High or low flue gas temperature
f. Frequent on-off burner cycle
g. Low condensate temperature
h. Water leaks through water feeder
i. Leaking relief valve
j. High gas pressure
k. Combustion analysis

3. Energy Waste Solutions

   a. Soot slows down heat between flue gases and water. Brush tubes clean.

   b. Mud inside bottom indicates contaminated water and lime on water side of tubes. This slows down heat to water. Blowdown water daily, until boiler water clears. When boiler is drained, wash out bottom with hose. Look for corrosion.

   c. Leaks mean cold water make-up using more fuel.

   d. Wet insulation increases heat loss from outside surface. Dry out or replace.

   e. High flue gas temperature – above 450°F for gas and oil – means over firing; or could be poor heat absorption by boiler surfaces (soot or lime). Low temperatures – below 300°F when burner is on – could mean excessive draft leaking air into boiler. Check for air leaks at base and between sections on cast iron boilers. Blow cigarette smoke, or use smoke tube.

   f. Quick on-off cycle indicates boiler may be too big for load. Readjust burner at lower capacity if full capacity not needed on start-up. If burner is modulating, limit the maximum burner condition. Check with the burner manufacturer for methods of this control.

   g. Low condensate temperature (below 180°F) means more fuel is needed to heat water to boiler temperature. Check make-up water into condensate tank, or poor insulation on return lines.

   h. If water feeder leaks through, water level will build up and you will have to blow down more often. This means you are heating up cold fresh water and throwing away hot water. Feel city water pipe; if it is colder than the room temperature, city water may be coming through.

i. Leaking relief valves waste steam or hot water. Requires water make-up and uses more fuel.

j. Check gas pressure; approximately 4 ounces for gravity burners, 4 to 15 ounces for power burners. Excess pressure wastes gas, since it may not all burn. High carbon monoxide in flue gases will indicate gas waste.

k. Run chemical analysis by sampling flue gas between boiler and draft diverter or draft regulator. Gas should be 80% to 82% efficiency. Oil: 80% to 83% efficiency. High oxygen readings indicate excess air. Check burner damper and draft in boiler. Boiler draft should not exceed .10 in. water gauge negative for gravity burners; .15 in. water gauge positive for pressurized boiler combustion.

Blow smoke into burner air intake or at front door when burner is not running. High draft due to high chimneys or induced draft fans will cool off boiler when shut down. Provide draft regulator or stack damper to stop suction of air at burner shut down.

4. Burner Operation

a. On-Off Cycles

If burner short cycles, check the range on the pressure stat for steam, or aquastat for hot water. Unless heating system is adversely affected, steam pressure range could be 3 pounds on low pressure and 10 pounds on higher pressure. Hot water temperature range can be 10°F. Short cycling wastes fuel.

b. High-Low Cycles

Time the burner operation at low fire and high fire. Adjust the maximum combustion efficiency for the firing level of the greatest time period. Test burner at both firing levels. Many burner controls require different air intake damper position and fuel pressures for varying fuel feeds. Check also for control span.

c. Modulating

Check control span as for on-off and high-low burners. As heat demand reduces due to season, provide a limiting device to prevent burner from going to full fire on start of heat cycle. This will save fuel and make the burner last longer.

5. Boiler-Burner Ratings

Relation of heat absorbing surface in a boiler to the heat output of the burner varies considerably. It is generally accepted today that a steel fire tube boiler should have 5 sq. ft. of heating surface per boiler H.P. (30,000 BTU/hr.).

3. Impurities in Fuels

a. Water

When water is heated above 212°F, it turns to steam. This absorbs 1000 BTU of heat for every pound of water heated. This uses more fuel.

b. Sulfur

When sulfur burns, sulfur dioxide results. Some sulfur trioxide also forms. This forms sulfuric acid, with water, and causes corrosion. If flue gases get below 265°F, with sulfur, corrosion will occur. Some heavy oils from South America contain high percentages of sulfur. Look at your fuel analysis. Some coals contain as much as 8% sulfur.

4. Heating Value of Fuels

BTU = British Thermal Unit = Heat to raise the temperature of one pound of water one degree Fahrenheit.

Natural Gas – 1,000 to 1,030 BTU per cu. ft.
Propane   – 88,217 BTU per Gallon, average
Fuel Oil #2 – 138690 "   "   "   "
     #5 – 146800 "   "   "   "
     #6 – 149690 "   "   "   "
Coal   – 9,000 to 14,500 BTU per pound

5. Incinerators

When paper, plastics, garbage and other waste materials are burned, the flue gases are the same, except there are a lot of impurities. The impurities cause odors, if not completely disintegrated. Plastics like polyvinyl chloride generate hydrochloric acid, which also corrodes metals.

Boilers

1. <u>Cast Iron Sectional</u>

   a. Gravity Gas

      In this type, the hot gases rise vertically to the top and are removed through a draft diverter to the flue. The open draft diverter prevents the draft in the boiler from rising much above zero.

   b. Combustion Chamber Type for Power Gas or Oil Burners

      Hot gases go to the back, rise up and pass forward through horizontal flues and then back again to the vent. These Boilers must have joints between sections sealed tight. Combustion Chamber should operate at about one tenth inch of water, negative pressure.

   c. Maintenance and Operation

      (1) Keep return headers clean. Blow down daily, if steam; once per week to clear water for hot water. Use blowdown oftener as water analysis indicates.

      (2) Check for air leaks between sections, or where cast iron meets the base.

      (3) Keep the refractory over the bottom legs of the sections to prevent burnout.

      (4) Do not let flame impinge on back wall.

      (5) Do not operate at pressures higher than rating.

      (6) With hot water, circulating pump must flow water through boiler before burner comes on. Do not fire burner without this water flow. Cast iron may crack.

      (7) Clean soot from flues

      (8) Keep insulation intact and dry

2. <u>Fire Tube</u>

   a. High Fire Box – Fire in Tubes, Water in Tank

      (1) This boiler can be fired with gas, oil, or coal It sets on a refractory lined base and floor. Flame expands in the combustion chamber, rises up in back and comes to the front through steel tubes sealed into tube sheets by rolling or

welding. Gases may go up to vent in front or go back to a flue in the rear, through another set of tubes. Water surrounds the tubes and extends down the sides, back and front of the combustion chamber to the base.

(2) Maintenance and Operation

All items listed for Cast Iron Boilers apply to this boiler.

b. Package Water Base Type (Scotch Marine)

(1) Description: This boiler has a steel tube combustion chamber, completely surrounded by water. This permits operation at high flame temperature, and under positive flue gas pressure since the entire boiler is sealed. Flue gases rise up in the rear and come forward through tubes to the front and up to the vent on a two-pass, to the back again on a three-pass, and again to the front on a four-pass. The back of the Combustion Chamber may be refractory lined, or have a water cooled panel called a "wet back." Operating pressure for steam or hot water can be increased, depending on the rating.

(2) Maintenance and Operation

(a) Blow down at least daily on steam and weekly on hot water, depending on water analysis or the amount of make-up water. This keeps the bottom of the Combustion Chamber from failing.

(b) Do not let the burner flame impinge on rear wall. This burns out refractory or warps back panel on wet backs.

(c) Keep flues clean

(d) On hot water, do not operate burner unless there is pumped water flow through boiler. If not, "thermal shock" may occur, which may snap the tubes in two at the tube sheet.

(e) Keep insulation dry and intact.

(f) Do not operate at pressures above rating.

(g) Make sure there is enough draft to remove flue gases from boiler. Excess pressure may cause gases to be forced out the front, not only making the burner inefficient but possibly burning out the front wall.

3. <u>Water Tube</u>

   a. Description

      (1) In these boilers, the water is in the tubes and the flame and flue gases pass over them. Return water comes into a tank or header at bottom. Steam or hot water rises to a tank or drum at the top. It is critical that water must be in all tubes exposed to flame, or the tube metal will fail by overheating or cracking.

      (2) In large boilers, walls may be covered with tubes, also called "water walls", to absorb more heat.

      (3) These boilers may be rated at very high pressures, depending on construction.

      (4) In small sizes, burner may fire right on tubes with a small refractory chamber, with oil, or direct with gas.

      (5) In large sizes, floor may be refractory. Some package units have tubes all around the Combustion Chamber, including the bottom.

      (6) Steam can be super heated, by taking the steam from the upper drum through another set of tubes exposed to hot flue gases. Water Tube boilers can be gas, oil, or coal fired.

   b. Maintenance and Operation

      (1) Blow down at least once per eight hour shift on large steam and once per week on hot water, depending on water analysis. Continuous blow down should be used if there is a lot of fresh water make-up Continuous blow down at water level in steam drum is required if oil, foaming or surging is present. Clean and open tubes are absolutely mandatory to prevent tube failure.

      (2) Keep tubes clean on outside, with soot blowers, if oil or coal fired.

      (3) Check for air leaks into the boiler; through walls, through base at floor and around doors. Excess air cools off boiler.

      (4) Inspect refractory floor, walls, bridge wall, and front panel around burner for damage. Repair before burn out. Shape of burner arch is critical to efficiency. Accumulation of

carbon on floor (on oil fired) and hot spots (on gas fired) also indicates poor burner flame pattern and low efficiency.

(5) Do not let flame impinge directly on tube bundle or water walls, unless there is sufficient water flow in tubes to protect them.

(6) Because of fast steaming, continuous water feed is essential. Feed water pumps and regulators must be kept in top condition.

(7) Do not operate at pressures above rating.

(8) Keep outside insulation, brickwork, steel jacket, or plaster, in good condition and dry.

(9) Keep draft at a minimum of one tenth inch water pressure in fire box, in standard construction, and at not more than one quarter inch positive air pressure on welded jackets.

(10) If floor is exposed in a basement, spot check surface of under side for temperature. Over 400°F may indicate insulation failure.

4. General Maintenance Items for All Boilers

a. Fix all water leaks, especially at Man Holes, Hand Holes and Plugs.

b. Blow Down all water feeders, low water cutoffs, feed water regulators, relief valves, gauge connections, pressure controller connections, at least daily, or in accordance with experience on sludge accumulation.

c. Test low water cutoffs and pressure controllers for operation at least once a week.

d. If you have a steam flow meter, of the orifice or venturi type, make sure the pressure tubes are open.

e. Clean out draft and flue gas pressure taps, depending on the cleanliness of the flue gases.

Fuel Burners

1. General

a. Fuel burners, whether gas, oil, coal or rubbish, should supply fuel and air in the proper proportion to burn completely and to supply the amount of heat to produce the steam or hot water required. Too much fuel feed wastes this fuel out the vent. Too much air feed cools off the fire and uses more fuel than necessary.

2. Types of Burners

    a. Gas Gravity Burners

        (1) Description

            (a) Gas, at pressure of about four ounces, is fed into a hollow casting with a lot of drilled holes in the top. As the gas comes out of the jet in the mixer, it draws in air. This mixture comes out of the holes and is lit by a pilot light. The flame also draws air from underneath the burner, to improve combustion.

        (2) Operation

            (a) Gas pressure and flue draft are critical. Higher gas pressure than rating will waste gas which hasn't been burned completely. If there is too much draft, more air than is needed will be drawn in. This may be more than is necessary to burn the gas, thus requiring more gas. This is the reason the draft diverter on the vent is wide open.

    b. Gas Power Burners

        (1) Description

            (a) In these burners, a blower draws in air through an adjustable intake. This air then mixes with,the gas under pressure in a mixing chamber. This mixture blows out of nozzles in a ring, through holes in a refractory, or sprayed out over a target. A pilot then lights the gas-air mixture. The shape and size of the flame should be adapted to the boiler combustion chamber. Flame can be long and narrow to short and wide. Controls can operate the burner on-off, low fire-high fire, or modulating.

        (2) Operation

            (a) On on-off burners, air intake damper can be set to supply the right amount of air with the gas feed. Gas pressure regulator must be fixed. Combustion test should indicate best combination for about 80% efficiency.

(3) Operation Limits

(a) On high-low fire and modulating burners, motor control on air intake damper should be set for top efficiency, while measuring combustion. Since efficiency changes with capacity, the best setting should be set at highest fire. If burner runs for long periods of time, at an intermediate position, the highest efficiency should be set at that capacity. Make sure the damper linkage is free and has tight connections.

When burner is off, air damper should close tight. If not, air may be sucked in by stack draft, cooling off the boiler.

(4) General Adjustments

(a) Keep gas pilots adjusted to right length to ignite flame.

(b) Keep safety pilot clean and adjusted, both thermostatic and flameye.

(c) Watch for changes in gas pressure. Use a gauge on the incoming line and after the control valve. Check daily. If main pressure gets too low, in cold weather adjust regulator. Don't forget to readjust it when main line pressure returns.

(d) Flue gas temperatures should vary from 300° to 450°. Lower temperatures indicate too much air. Higher temperatures, too big a flame for boiler capacity.

c. Pressure Atomising Oil Burners (Gun Type)

(1) Description

(a) This type of oil burner uses oil pressure at about 125 pounds per sq. in. The oil is forced through a small hole in a nozzle. This oil spray mixes with a stream of air, surrounding the nozzle, in a tube pushed in by a blower. Mixture is lit by an electric arc, between two high voltage electrodes.

(2) Operation

This burner is used on small sizes, with No. 2 or No. 3 oil; on-off control. Outside air can be fixed.

(3) Maintenance includes keeping nozzle clean, adjusting ignition electrodes, checking oil pressure. If flame goes out periodically, look for water in oil or air leaks into the suction line from the tank. Watch the pressure gauge, while operating, for fluctuating pressure. Low pressure may indicate worn oil pump. Clean flame rod detector. Other problems similar to gas burners.

d. Steam or Air Atomising Oil Burners

(1) Description

This burner uses pressures of 100 pounds per sq. in. to atomise the oil through nozzles, usually with a target disc or cone. Oil is forced out in a swirling motion, to mix oil particles and air to produce a spreading flame.

(2) Operation

If steam is continuously available, this can be used for atomising. If not, air is supplied by a separate compressor. Oil pressure should be monitored also. Capacity control by adjustment of both oil and air, in proportion to the heating requirements, uses oil valves and air dampers on blowers. Adjustment of both oil and air should be set while measuring combustion efficiency at 100% operation. Efficiency will reduce on light load operation.

(3) Maintenance

Since burner is large, a gas flame or continuous flame is usually used as a pilot. Some electric pilots are available. Maintenance problems are similar to pressure atomising burners.

e. Rotary Cup Oil Burners

(1) Description

In this type, oil is fed onto a rotating cup which flings the oil droplets in a circle. Air is blown around the cup, forming again in a swirling flame. Cup rotation speed is constant.

(2) Operation and Maintenance

Control, operation and maintenance are similar to other oil burners. Cup must be kept clean, carbon removed and shape and surface kept the same as new.

  f. Coal Burners

    (1) Description

There are a number of types of coal burners: Under Feed, Spreader, Chain Grate, Pulverized. Air and fuel proportion are also necessary; but are limited by ash quantity, ash melting temperature and fuel quality.

    (2) Operation

Coal beds cannot be turned off, so controls must feed air and fuel at low rate to keep flame from going out.

    (3) Maintenance

Because of ash, boiler must be cleaned out and tubes soot blown often. Ash removal and grate repair is essential.

Boiler and Burner Controls

1. Operating Controls

  a. Steam Pressure

Steam pressure is measured by pressure control in the header or in the steam section of the boiler, which will operate the burner within the selected pressure range. There is an additional high pressure switch, used for safety purposes, which will shut off the burner in case the controller does not function. In the case of modulating burners, this control positions the fuel flow and the combustion air intake damper in proportion to the requirements of the boiler. This type of control can keep the burner on continuously, at a reduced capacity, so that it has an even steam pressure. On/off burners will vary, usually in the range of approximately three pounds, from start to stop on low pressure boilers up to fifteen pounds; and on high pressure boilers at 100 pounds or more, this variation may be increased to approximately 10 pounds.

Compare the operation of the pressure stat or switch with the pressure gauge to determine if there is variation in its calibration.

On steam boilers, if it is required to have a rapid pick up from a cold boiler, it is recommended that a hot water stat be installed, below the water line, which will keep the burner operating and keep the water temperature just under the steam temperature at all times. Then, when you are ready to produce steam, the water has already been heated up to almost the boiling point and steam will be produced quickly.

b. Water Temperature

On hot water systems, a thermostat in the supply hot water line, or in the boiler itself near the top, will control the burner in the same manner; that the steam pressure stat does. Range of temperatures is in the neighborhood of 5 degree fluctuation from on to off and modulating thermostats and burners are used for hot water also. Here again, check the operation of the temperature stat with the thermometer, located in the same line as close as possible to this instrument, to check its calibration.

c. Gas Feed

Natural gas, or vaporized propane, is fed into the boiler burner through a pressure regulator to two automatic control valves, one of which is open and shut for safety and the other may be on/off, or modulating, depending on the type of burner. There is a vent from the top of the pressure regulator and also from the line between the two control valves, as required by code. A very small amount of gas leakage should be run into the boiler and be burned off at the flame or allowed to dissipate up the stack. In addition, there is a pilot safety which prevents the valve from opening if the flame does not light, or there is insufficient pilot flame or control for this purpose. These safety pilots are described under safety controls. A damper on the air supply should be proportioned directly with the gas feed, so that the proper amount of air for combustion is supplied. Check the requirements of the burner, as to gas pressure, since certain burners work at approximately 4 ounces of pressure and may become inefficient, or dangerous to operate, if the pressure entering the burner is greater than the rated amount.

d. Oil Feed

Oil is supplied to the burner by means of the pump which pressurizes the oil for the purpose of forcing it through the nozzle, or small opening, in the feed pipe. There is also a control valve which may open and close, or may modulate in the same manner as gas burners. These valves also are operated by the same type of pressure or temperature control as gas. Oil is ignited, usually, by a gas flame or by electric spark. Heat can be shut down by flame failure control.

e. Air Feed

Combustion air is supplied to the burner by means of a blower, in the case of pressurized burners, and by means of the jet action of the gas through an orifice on gravity gas burners. The same steam or pressure water temperature control operating the fuel flow also operates the motor controlling the position of the damper on the air intake to the burner to provide combustion air in the proper ratio. Some of these damper controls are electric on/off or proportioning type tied into the control circuits for the fuel flow and some of them are direct mechanical devices, using a shaft drive with an arc positioning determining device to set the air intake damper.

f. Water Feed

In steam boilers, condensate coming back to the boiler is fed by means of a pump receiving the water from a collecting tank. Additional water, if it is not lost or if it settles out in parts of the system, is supplied through a float type water feeder connected to the side of the boiler at the water level. If the water level goes below a recommended fixed position, a float will open a line directly connected to the city water or domestic water source, to refill the boiler up to the proper operating level. This level control can also be an electrode probe type, operating on the conductivity of the water, and which, in turn, will open an electric valve on the water makeup line and refill the boiler. On hot water boilers, a similar float control is used for safety purposes. The water is connected directly to the system at all times, through a pressure reducing valve which keeps a constant minimum water pressure on the system. An expansion tank, connected to the hot water supply main with an air cushion, permits the expansion and contraction of the water in the system, due to temperature changes, to prevent excess water feed to the system when water is cold and to prevent loss of water from the system when water is hot. These expansion tanks should have an air vent, drain cock and gauge glass on the side so they can be inspected occasionally to find out if they are completely water logged.

g. Draft

Draft control is a sensitive pressure switch, operating at very low pressures, in the range of 0 to 5 inches of water pressure and connected to the stack or vent. This control operates a damper in the stack, which may reduce the suction of the stack under light loads. The purpose of this control is to prevent too great a suction on the windbox and the combustion chamber, which will draw in more air than required for combustion and thus make this burner quite inefficient. These are particularly useful on high stacks in cold weather, when the effect of draft due to cold outside air and warm air in the stack creates an excessive suction on the boiler.

h. Steam Flow

Steam flow is usually measured by an orifice or venturi flow meter and normally does not control the burners. The dropping of pressure in the boiler indicates that higher steam flow is occurring and the pressure then operates the burner.

i. Water Flow

Water flow is measured also by orifice meter or by displacement type meter, but does not control the burner. Higher water flows are indicated in a drop in temperature, which, in turn, turns the burner on. To safeguard the possibility of extreme reduction in water flow through the boiler when the boiler is hot, it is recommended that a recirculating pump be installed for minimum water flow through the boiler so that the boiler will not crack in the case of cast iron, or rupture the tubes in the case of a fire tube or water tube boiler. This pump would only operate when the water flow slows down too greatly. This control is in the form of a paddle type flow switch which will shut off the burner or turn on the recirculating pump at low water flows.

j. Fuel Pressure

Although not normally used, fuel pressure switches should be installed in the system to prevent the gas burners operating at too high a gas pressure and the oil burners operating at too low an oil pressure. If there is possibility of considerable fluctuation and pressures on these two fuels, consider the possibility of adding a safety pressure control on each.

2. Safety Controls

a. Pressure Relief

Both steam and hot water boilers, and certain types of heat exchangers, must have a means of preventing pressure buildup higher than the equipment is designed, Pressure relief valves provide this safety. These valves must be tripped occasionally, to make sure they work. Failure to work may cause the boiler to explode or the outside to rupture.

b. Temperature Relief

On hot water boilers, the relief valve shall also work on temperature. Because of possibility of internal steam generation, this could cause internal tube rupture.

c. Gas Pressure

Gas pressure relief valves or bleeds should be used on the leaving side of pressure regulating valves. Excess gas pressure can cause raw gas to go up into the flues, where fire or explosion can occur. Excess pressure wastes gas also, because the burner does not supply enough air to burn off all the gas. Gas relief should be to the outside or the flue.

d. Oil Pressure

Oil pressure relief is required for the same reason as gas. Relief should be back into the pump intake or to the return line to the tank.

e. Flame Failure

If the fuel mixture does not ignite, the burner must be shut down to prevent filling the combustion chamber with explosive fumes. If the boiler is hot, an explosion usually occurs.

Flame failure devices work on a thermostat for gas. The pilot flame on the thermocouple element keeps the automatic valve open. On oil, a rod in the pilot flame, or photocell looking at the flame, is the flame failure device.

f. Water Level

Surfaces exposed to flame or hot flue gases must be protected with a backup of water or steam, to keep the metal from failing. Water level safety controls are float type or electronic probes. The float type must be kept clean, so float will be free. Blow down often. Probe type must have clean electrodes. Remove and inspect at least once a year and remove scale and lime.

g. Stop Check Valves

On multiple boiler installations, where boilers are equipped with manholes, these valves prevent steam from entering a shut down boiler. This is for safety of anyone working inside the cold boiler.

h. Explosion Relief

If there is a restriction in the breeching or stack, like an economizer, there should be a weighted relief damper connected to the combustion chamber in case of pre-ignition or "puff."

3. Environmental Control

a. Opacity

In the case of oil firing, and especially with coal firing, the E.P.A. requires some means of measuring smoke generation or concentration of solid impurities in the flue gases, by means of opacity control. This device is usually a light beam, projected across the stack through a series of mirrors. This device, depending on the requirements of E.P.A., will either sound an alarm or will shut down the burner.

b. Oxygen Control

Although oxygen is not an environmental problem, the measurement of oxygen content is important from the standpoint of measuring the excess air and the efficiency of the burner. Since a lack of oxygen indicates the possibility of insufficient burning of the fuel, generating carbon monoxide and, in some cases, raw fuel in the flue gases, this is an important device, not only from saving energy but from safety and pollution. These devices use a zirconium oxide sensing unit which measures the oxygen content in the flue gases. These devices also require a continuous cleaning operation to make them effective, by the use of an air supply or sometimes with a water wash.

4. Efficiency Controls

   a. Oxygen Trim

   The same oxygen measuring device as listed under Environmental Controls is used to operate the combustion air damper or the combustion air supply, to provide the correct amount of excess air for combustion. Although combustion efficiency is sometimes measured in the form of carbon dioxide, there is not a good device made for measuring carbon dioxide continuously and, since oxygen is more important as a measuring device, this equipment is quite successful and should be used.

   b. Draft

   Draft is essential to be measured and uses a very sensitive pressure switch to control dampers, usually in the stack. This is described under item 1-g.

5. Instruments

   a. Operating Instruments

      (1) Sensitive Pressure Gauges measure draft in chimney or vent; draft or pressure in the firebox; and burner blower pressure.

      These gauges measure small pressures in inches of water column. Clean sensing pipes and tappings often, if there is dirt or soot present.

      Too much draft in firebox may draw too much excess air; insufficient draft in vent may prevent removal of flue gases.

      (2) Steam Flow

      Steam flow meters are usually a flat plate with a round hole, mounted in a pair of flanges. This is called an orifice meter. It measures steam flow by creating a pressure differential across the plate. This is then transmitted to a recorder. In the recorder, bellows like in a pressure gauge, move the recorder pen. The faster the steam flow, the greater the pressure difference. It is recommended that the orifice plate be removed at least once every five years. Check the hole for size, shape and smoothness. The author has tested boilers whose orifices were off 20%.

Another meter type is a venturi, a sloping tube which works the same as the orifice.

There are some new devices available, using a pitot tube or a paddle wheel.

(3) Hot Water Flow

Water, hot or cold, can be measured also by the office meter or with a displacement or turbine meter like the one in your house.

The pitot tube and paddle wheel can be used here also.

(4) Gas Flow

Natural gas uses a diaphram type meter for low pressure. On pressures of one pound or higher, a rotameter, a glass tube with a floating disk, can be used for indicating. Orifice meters similar to that used for steam give a similar recording. For large volumes, a turbine or roots cycloidal meter is used.

(5) Oil Flow

Oil flow is measured by a displacement or turbine meter like that for water. Since most oil pumps are gear type, if you know how much oil is between the gear teeth, you can figure the oil flow from the pump R.P.M.

(6) Pressure

Pressure of both steam and hot water is indicated by gauges using an expanding element to move a needle on a dial. This element can also operate an electronic device to transmit a signal to a remote indicator.

Check these gauges periodically, with a calibrated master gauge, as vibration and moisture inside operating tube may change reading.

(7) Temperature

Temperatures, measured by thermometers which may be filled with a liquid such as mercury or a bimetallic,element which expands and moves the indicating needle. Thermo-couples are also used which are voltage generating devices, operating an electric milivoltmeter. Thermo-couples are used with higher temperatures; the direct reading type at lower temperatures.

(8) Recorders

Recorders are used to measure steam flow, water flow, temperature and pressure; are similar to the indicating devices, except the needle is now a pin on a rotating chart, indicating a condition at a certain time of day. These recorders should be used on all fuel burning equipment to determine the characteristics of use of steam or hot water and whether there is any change in the consumption of the system. Excessive operation may indicate leaks in the system, or inefficient equipment.

b. Field Testing Instruments

(1) Temperature; see pare. E.Sa(7) and Fig. H.1

(2) Pressure; see pare. E.5a(6) and Fig. E.5a(6)

(3) Steam Flow; see pare. E.5a(2) and Fig. E.5b(3)

(4) Water Flow; see pare. E.5a(3) and Fig. E.5b(3)

(5) Combustion

Combustion is measured by means of liquid absorption equipment, for both carbon dioxide and oxygen. This is a dumbbell shaped device in which the flue gases are pumped into the liquid, the device shaken and the carbon dioxide is absorbed by a chemical which causes the level to change. The change in the level indicates the percentage of carbon dioxide. The same type of liquid device is used for measuring oxygen, with an oxygen absorbing solution. Carbon monoxide is measured also by a liquid, or by a glass tube which discolors over a length of the glass tube. This length of discoloration is measured and gives a percentage of parts per million of carbon monoxide.

A better device is an electronic unit, similar to the oxygen measuring instrument as above mentioned, using a sensor which creates an electrical circuit if in the presence of a certain percentage of oxygen. These devices are much more accurate than the liquid units and can run continuously by means of a vacuum pump which pulls the flue gases through the device and measures the oxygen content. This particular instrument should be protected from the admission of moisture, as this will damage the instrument. The same instrument will measure hydrocarbons which indicates that there is unburned fuel, or carbon monoxide in the flue gases, thus determining improper combustion.

(6) Draft

Draft is measured by means of very sensitive pressure gauges, measuring pressure as low as 1/5 of an inch of water column, and are necessary for proper adjustment of the amount of air pressure within a combustion chamber, created either by the burner and its air supply or by the difference in the temperature and the height of the stack. If these devices should read both negative and positive pressures, as draft is a negative pressure and some boilers operate under positive pressure conditions.

(7) Air Flow

Combustion air flow, or flue gas flow, can be measured by a portable vane or electronic meter. For measurement in ducts, or at higher temperatures, a pitot tube is used. This measures the velocity pressure versus the static pressure. The difference measures the flow. Keep the nozzles and sensing ports clean.

(8) Flue Gas Flow

Use high temperature pitot tube.

(9) Electric

Electric current is measured by an ammeter; voltage by a voltmeter; and watts by a wattmeter. If you know the conversion figure on the utility electric meter, you can read kilowatt hours, if the meter supplies only the heating system.

Voltage and amperage of electric motors and electric devices should be measured by means of a combination of voltmeters and ammeters. On alternating current lines, this device can clamp around a wire and measure electric current and also in contact with two of each of the three legs of three phase current, can indicate the voltage on the line. Check the running current rating on the motors and on the devices as voltage and amperage determine if the equipment is overloaded or underloaded.

(10) Rotational Speed

RPM is measured by a tachometer, in contact with the shaft, or an optical device which flashes light at the same speed. A mark is needed on the rotating device to measure optically. You should also know about what speed the shaft should be, because the light stops the action at multiples of the original speed (such as 1750 RPM, 3500 RPM, 7000 RPM).

Boiler Plant Accessories

1. Feed Water Systems

    a. Feed Water Pumps – Centrifugal, Turbine

    On Feed Systems, water must be fed back into the boiler, at a pressure above the operating pressure, or the water will not enter the boiler. Pump is of two general types: Centrifugal and Turbine. The centrifugal pump operates by means of water sliding off the edge of a backward curved, bladed impeller and is non-overloading. When it is overloaded, the water slips past these blades. Normally, if the centrifugal pump is run for a period of time with very little water flow and at maximum pressure, the water will heat up to steam, due to the input of the motor, and burn out the seals.

This is not true of the Turbine type pump, which is normally used in small sizes at reasonably higher pressures. The Turbine pump has very small openings and is a flat blade unit and will overload. If the water flow is reduced, or the pressure becomes too high, on a turbine pump, the pump will overload. Check to see which type of pump you have on your feed water system.

In addition, there must be sufficient positive pressure on the suction side of the pump to prevent the water from flashing into steam and vapor binding the pumps. This is called N.P.S.H. Look at the type of pump you have and find out if the water temperature is such that the water entering the pump may flash. This is the reason that the condensate receivers, or storage tanks, are placed above the pump suction – to promote the pump to stay primed and not cause this difficulty. If the pump chatters, this may be a condition of such flashing. Under high pressure conditions, the pump may be multiple stage to create this pressure and, under these higher pressure conditions, the boiler should be supplied with a pump controller type relief valve, which puts the water back into the suction side when too great a pressure is generated due to reduced water flow or due to the operation of the feed water controller.

b. Feed Water Heaters

Most systems which lose water, in the form of condensing steam in the process, or are loosing water and require considerable makeup, the condensate and feed water should be heated to at least 225° or the equivalent of 5 pounds steam, to drive the oxygen off. If this is not done, especially on water tube type boilers, considerable corrosion will occur. These heaters are usually either spray or tray type, in which the water extends down through the receiver tank and steam is passed vertically through this water stream to drive the oxygen off. The oxygen is then vented to the atmosphere. It is essential that the temperature of this water, and its equivalent pressure, do not exceed 225° because the water will flash into steam and considerable steam will be lost out the vent.

  c. Condensate Pumps

   The return water from the steam system comes back in the form
   of condensate and usually ends up in the receiver. The pumps
   pump this condensate back into the boiler in the smaller systems, at a pressure above the operating pressure in the boiler.
   These pumps can be operated from a float control on the boiler,
   intermittently, or may be operated from a float control within
   the receiver tank, to keep the level of the water in the receiver
   tank constant. The continuous feed control on large boilers, is
   more efficient as this provides a constant level of water and a
   constant volume of steam within the boiler itself.

  d. Condensate and Feed Water Storage

   These tanks should be sized in proportion to the amount of condensate and feed water used in the system, so that there is no
   overflow and loss of water. These tanks should be insulated so
   that the condensate does not lose heat while being stored.

  e. Turbines

   On steam systems exceeding 100 pounds and preferably those
   in the range of 250 pounds and higher, steam can be used to
   drive turbines for the pumps instead of using electric motors.
   Use of turbines should be a function of the possibility of using
   the exhaust steam pressures at 5 to 10 pounds. Turbines are single and multiple stage and can operate with discharge pressures
   at approximately 1/3 the supply pressure; or, if full capacity is
   desired, can discharge the exhaust into a condenser which creates a vacuum and then the condensate thus generated is
   pumped back into the feed water system.

2. Blow Down Systems

  a. Valves

   Steam and hot water boilers must be blown down occasionally
   to remove the concentration of impurities which will affect the
   heat transfer and the corrosion within the equipment. These
   valves are a special type, usually operating on a 45 degree
   angle, so that the impurities will wash off the seat and not prevent it from tight closing. These valves should be inspected to

determine if they are leaking, since leaking condensate and hot water is expensive.

b. Continuous

On large systems, especially of the water tube type, the blow down should be continuous so that not only the so-called Mud Drum, but also from the surface of the water in the steam drum where oil or foaming materials may occur. When this is done, a metering device is required to prevent excessive blow down.

c. Water Cooling

Since the blow down water is too hot for normal sewer use, most communities require a cooling of this blow down to a maximum of 140°F. This is performed by running the blow down through a heat exchanger to cool it off, or by mixing with cold water. The control of the cooling water is by means of a thermostat in the tank, operating a control valve on the water supply.

d. Economisers

Economisers are used in the flue gases to preheat the feed water, especially on the larger boilers where the flue gas temperature is in the range of 400 to 500 degrees. Do not use Economisers to reduce the temperature of the flue gases below 265 degrees, if there is any possibility of sulfur dioxide in the flue gases, because of corrosion. These Economisers are usually large tubes, with fins, and considerable spacing between them to prevent plugging up with soot. Check the draft pressure on both sides of the Economiser to determine whether they are starting to plug up with soot or fly ash in the flue.

e. Soot Blowers

If oil or coal is used, soot and fly ash will collect on tube surfaces and also on heat Economisers. The device necessary for removing this soot and fly ash is a soot blower, which is an air or steam pressure device blowing jets of air in a rotating method to remove this material. Make sure that there is some collecting device that will take the fly ash or soot out of the air stream, or do it under certain conditions that will comply with requirements of E.P.A.

Chapter 11

Heating Test Data Sheets

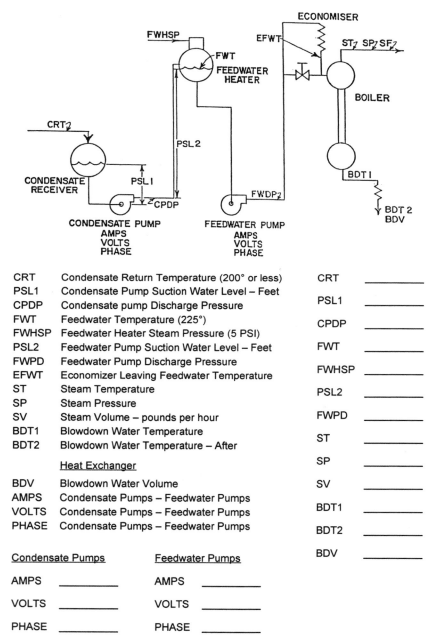

CRT   Condensate Return Temperature (200° or less)   CRT   _____
PSL1   Condensate Pump Suction Water Level – Feet
CPDP   Condensate pump Discharge Pressure   PSL1   _____
FWT   Feedwater Temperature (225°)   CPDP   _____
FWHSP   Feedwater Heater Steam Pressure (5 PSI)
PSL2   Feedwater Pump Suction Water Level – Feet   FWT   _____
FWPD   Feedwater Pump Discharge Pressure   FWHSP   _____
EFWT   Economizer Leaving Feedwater Temperature
ST   Steam Temperature   PSL2   _____
SP   Steam Pressure
SV   Steam Volume – pounds per hour   FWPD   _____
BDT1   Blowdown Water Temperature   ST   _____
BDT2   Blowdown Water Temperature – After
                                                             SP   _____
         Heat Exchanger
                                                             SV   _____
BDV   Blowdown Water Volume
AMPS   Condensate Pumps – Feedwater Pumps   BDT1   _____
VOLTS   Condensate Pumps – Feedwater Pumps
PHASE   Condensate Pumps – Feedwater Pumps   BDT2   _____

                                                             BDV   _____
Condensate Pumps   Feedwater Pumps

AMPS _____   AMPS _____

VOLTS _____   VOLTS _____

PHASE _____   PHASE _____

Figure 11-1 H1 / Steam Boiler Plant / Steam – Water System
Field Measurements

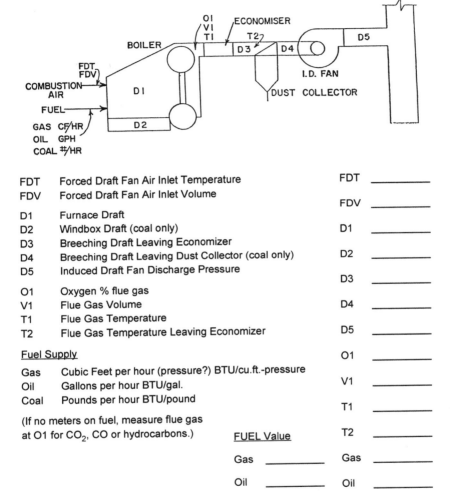

FDT	Forced Draft Fan Air Inlet Temperature
FDV	Forced Draft Fan Air Inlet Volume
D1	Furnace Draft
D2	Windbox Draft (coal only)
D3	Breeching Draft Leaving Economizer
D4	Breeching Draft Leaving Dust Collector (coal only)
D5	Induced Draft Fan Discharge Pressure
O1	Oxygen % flue gas
V1	Flue Gas Volume
T1	Flue Gas Temperature
T2	Flue Gas Temperature Leaving Economizer

FDT _____

FDV _____

D1 _____

D2 _____

D3 _____

D4 _____

D5 _____

O1 _____

V1 _____

T1 _____

T2 _____

Gas _____

Oil _____

Coal _____

CO2 _____

CO _____

Hyd. _____

<u>Fuel Supply</u>

Gas	Cubic Feet per hour (pressure?) BTU/cu.ft.-pressure
Oil	Gallons per hour BTU/gal.
Coal	Pounds per hour BTU/pound

(If no meters on fuel, measure flue gas
at O1 for CO_2, CO or hydrocarbons.)

<u>FUEL Value</u>

Gas _____

Oil _____

Coal _____

Figure 11-2 H2 / Steam Boiler Plant / Combustion System
Field Measurements

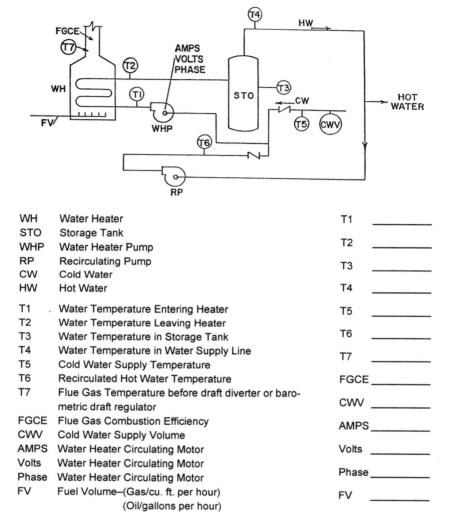

WH	Water Heater	T1	_____
STO	Storage Tank	T2	_____
WHP	Water Heater Pump		
RP	Recirculating Pump	T3	_____
CW	Cold Water		
HW	Hot Water	T4	_____

T1	. Water Temperature Entering Heater	T5	_____
T2	Water Temperature Leaving Heater		
T3	Water Temperature in Storage Tank	T6	_____
T4	Water Temperature in Water Supply Line		
T5	Cold Water Supply Temperature	T7	_____
T6	Recirculated Hot Water Temperature		
T7	Flue Gas Temperature before draft diverter or baro-	FGCE	_____
	metric draft regulator	CWV	_____
FGCE	Flue Gas Combustion Efficiency		
CWV	Cold Water Supply Volume	AMPS	_____
AMPS	Water Heater Circulating Motor		
Volts	Water Heater Circulating Motor	Volts	_____
Phase	Water Heater Circulating Motor		
FV	Fuel Volume–(Gas/cu. ft. per hour)	Phase	_____
	(Oil/gallons per hour)	FV	_____

Figure 11-3 H3 / Domestic Hot Water Heating / Heating Source Oil Gas
Field Measurements

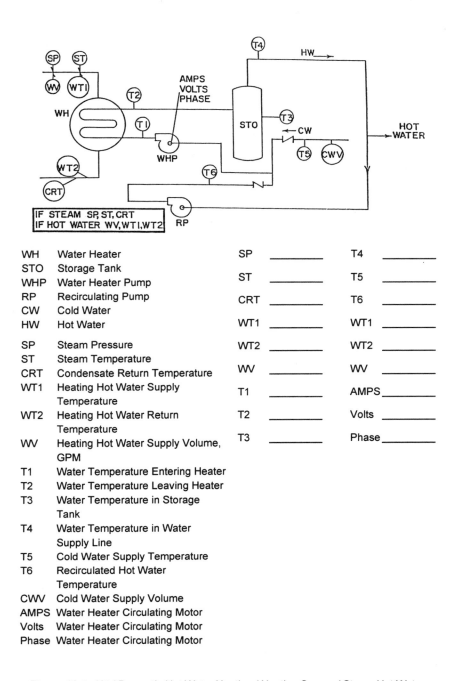

WH	Water Heater	SP _____	T4 _____
STO	Storage Tank	ST _____	T5 _____
WHP	Water Heater Pump		
RP	Recirculating Pump	CRT _____	T6 _____
CW	Cold Water		
HW	Hot Water	WT1 _____	WT1 _____
SP	Steam Pressure	WT2 _____	WT2 _____
ST	Steam Temperature		
CRT	Condensate Return Temperature	WV _____	WV _____
WT1	Heating Hot Water Supply Temperature	T1 _____	AMPS _____
WT2	Heating Hot Water Return Temperature	T2 _____	Volts _____
WV	Heating Hot Water Supply Volume, GPM	T3 _____	Phase _____

WH   Water Heater
STO   Storage Tank
WHP   Water Heater Pump
RP   Recirculating Pump
CW   Cold Water
HW   Hot Water

SP   Steam Pressure
ST   Steam Temperature
CRT   Condensate Return Temperature
WT1   Heating Hot Water Supply
          Temperature
WT2   Heating Hot Water Return
          Temperature
WV   Heating Hot Water Supply Volume,
          GPM
T1   Water Temperature Entering Heater
T2   Water Temperature Leaving Heater
T3   Water Temperature in Storage
          Tank
T4   Water Temperature in Water
          Supply Line
T5   Cold Water Supply Temperature
T6   Recirculated Hot Water
          Temperature
CWV   Cold Water Supply Volume
AMPS Water Heater Circulating Motor
Volts   Water Heater Circulating Motor
Phase Water Heater Circulating Motor

Figure 11-4 H4 / Domestic Hot Water Heating / Heating Source / Steam Hot Water
Field Measurements

Chapter 12

Illustrations

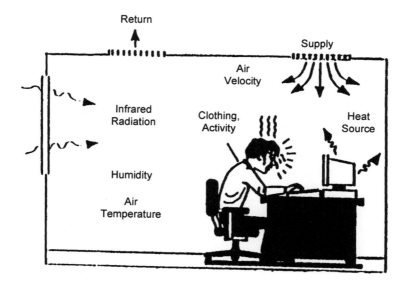

1. Comfort Temperature
   72° - 70% Relative Humidity
   74° - 60% Relative Humidity
   76° - 50% Relative Humidity

2. Body Heat Removal
   60% Radiation
   40% Evaporation

3. Air Motion
   1 Mile per Hour
   Ceiling Fans
   3 Miles per Hour

4. Cause of Comfort Variation
   What happened last night?
   What happened this morning?
   Clothing
   Physical activity
   Physical problems

5. External Source
   Sun
   Heat producing equipment
   Lights

Figure 12-1 Parameters impacting comfort in the office environment.
Office air velocity measurements should be between 20 and 50 fpm in occupied zones.
See page 2

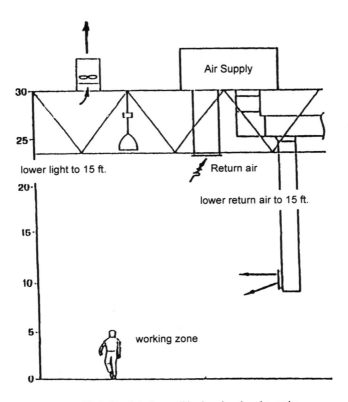

Figure 12-2 Partial air conditioning, low level supply.
Air stratification. See page 4

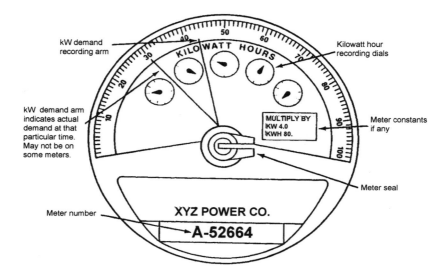

kW demand
recording arm

Kilowatt hour
recording dials

KILOWATT HOURS

kW demand arm
indicates actual
demand at that
particular time.
May not be on
some meters.

MULTIPLY BY
KW 4.0
KWH 80.

Meter constants
if any

Meter seal

Meter number

XYZ POWER CO.

A-52664

Figure 12-3 Typical Demand Meter

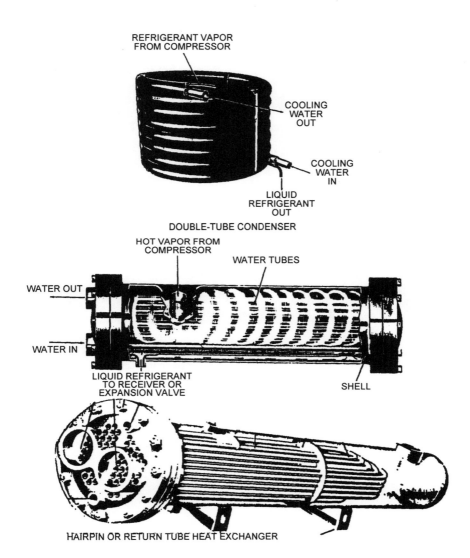

REFRIGERANT VAPOR
FROM COMPRESSOR

COOLING
WATER
OUT

COOLING
WATER
IN

LIQUID
REFRIGERANT
OUT

DOUBLE-TUBE CONDENSER

HOT VAPOR FROM
COMPRESSOR

WATER TUBES

WATER OUT

WATER IN

LIQUID REFRIGERANT
TO RECEIVER OR
EXPANSION VALVE

SHELL

HAIRPIN OR RETURN TUBE HEAT EXCHANGER

Figure 12-4 Water Cooled Condensers
See page 6

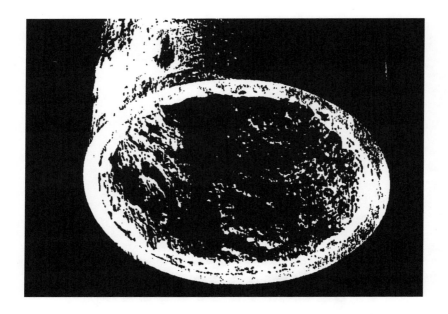

Figure 12-5   Scale deposits and corrosion products on tube surfaces
reduce heat-transfer efficiency and increase energy costs.
See page 6

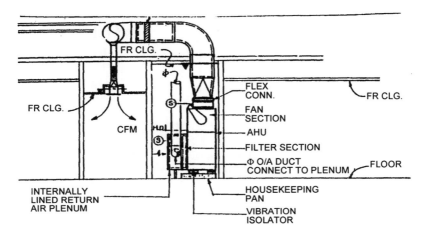

Figure 12-6 AC Unit Installation. See page 5

✓ Air-handling units should be constructed so that equipment maintenance personnel have easy and direct access to both heat exchange
components and drain pans for checking drainage and cleaning.
Access panels or doors should be installed where needed.

✓ Non-porous surfaces where moisture collection has promoted micro-
bial growth (e.g., drain pans, cooling coils) should be cleaned and
disinfected with detergents, chlorine-generating slimicides (bleach),
and/or proprietary biocides. Care should be taken to ensure that
these cleaners are removed before air-handling units are reactivated.

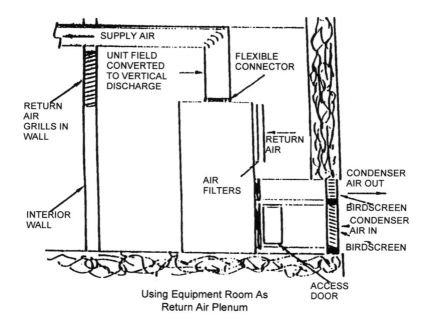

Using Equipment Room As
Return Air Plenum

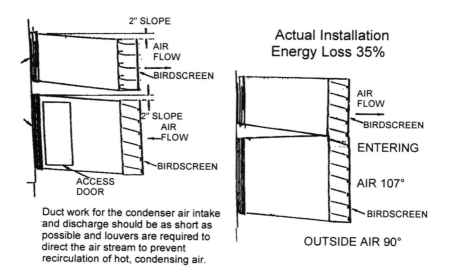

Duct work for the condenser air intake
and discharge should be as short as
possible and louvers are required to
direct the air stream to prevent
recirculation of hot, condensing air.

Figure 12-7   Recirculated Condenser Air. See page 11

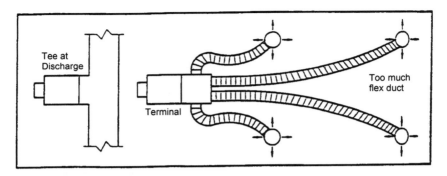

Figure 12-8 Flexible Duct.

FIBERGLAS DUCTS

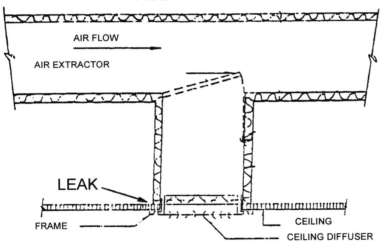

Figure 12-9 Typical Ceiling Outlet Detail (no scale)
See page 8

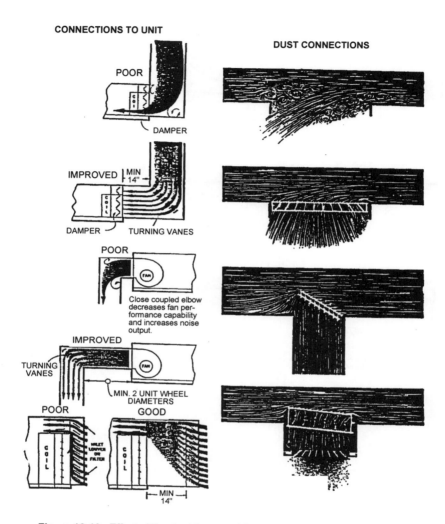

Figure 12-10   Effect of Turning Vanes and Deflectors on Air Flow (see page 8)

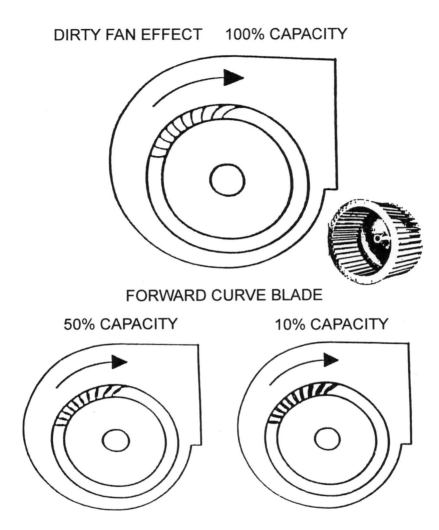

Figure 12-11 FC Fan (see page 8)

DESIGN AIR VOLUME

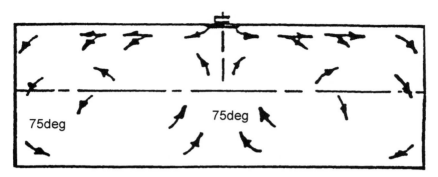

LOW AIR VOLUME

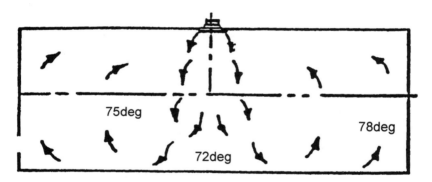

Figure 12-12   Ceiling Air Outlets (see page 9)

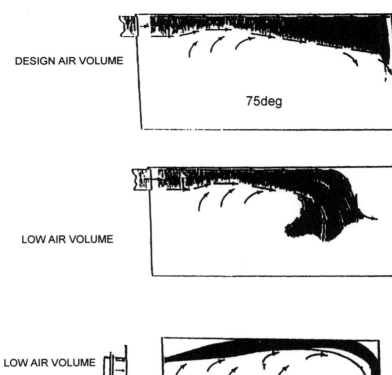

DESIGN AIR VOLUME

75deg

LOW AIR VOLUME

LOW AIR VOLUME

75deg

To correct adjust grille
horizontal blades 1/3 top down 1/3 bottom up

Figure 12-13   Wall Air Outlets (see page 9)

Chapter 13

Field Test Results

TEST OF TWO ROOFTOP AIR COOLED CHILLERS.

Building – 11 stories – 15 years old.

Chillers on 10th story roof section.

Chilled water pumps in 10th story equipment room.

Air system combination single zone, multizone, fan coil units.

Chilled water expansion tank on 11 story roof.

Location South Florida, 1/4 mile from Atlantic Ocean.

Test in April.

Outside air 80°F.

Building usage 9 month 100%, 3 months 50%.

Chilled water pumps.

    Discharge pressure.

        Left pump – 38.5 psi   Right – 33.5 psi

    Suction Pressure – 2 psi

Chilled water temperature.

        Leaving chiller – 45°

        Entering chiller – 52.5°

Pressure drop across pump discharge.

    Check Valve   Left – 0psi, Right – 2 psi

TEST RESULTS

Condenser Air Flow

	CFM Air Volume	Air Temperature In	Air Temperature Out
East Chiller			
Fan A	21074	85°	104.5°
B	16770	85°	105.7°
C	26254	85°	106.2°
West Chiller			
Fan A	24825	81°	100.2°
B	18360	83°	101.5°
C	21074	85°	102.1°

	East Unit	West Unit
Total Power Input	93.28 KW	103.8 KW
New Cooling Capacity	90.98 tons	68.6 Tons
KW Input/Ton	1.02	1.53

Specifications of Equipment

Chillers

102 Tons, 271 GPM, 51° ENT CH.W., 42° LVG CH.W., 10 FT H20 Pressure Drop, 95° ENT Air, 115° Condensing Temp., 3 – 7 1/2 HP Fans @ 11 AMP Each, Total Power Input 153 AMPS at 480 Volts.

Electric Readings

East Chiller	Specifications
Voltage 480, 478, 480	440
AMPs – Main – 115, 108, 114	153
Compressor – 102, 100, 88	120
Unloaded 60, 56, 63.8	
Condenser Fans	
A – 2.6, 2.3, 2.7	11.0
B – 4.4, 4.1, 4.4	11.0
C – 5.7, 5.5, 5.5	11.0

West Chiller	**Specifications**
Voltage 480, 478, 480	
AMPs – Main 155	153
Compressor 128, 123, 124	120
Unloaded 104, 96.5, 104	
Condenser Fans	
A – 4.5, 5.0, 5.7	11.0
B – 4.6, 4.2, 4.3	11.0
C – 4.0, 4.7, 4.4	11.0

Solution:

What is wrong with system?

1. Discharge air is recirculating.
2. Condenser fans slow.
3. Chilled water pressure low.
4. Chilled water temperature high.
5. Chiller water tubes coated.
6. Condenser coil dirty.

How to correct?

1. Put collars on condenser fan discharge.
2. Speed up condenser fans.
3. Raise chiller water pressure 22 PSI, Pump L, 56 PSI, R. 54PSI, Suct 24 PSI.
4. Drop chilled water to 42°.
5. Clean cooler tubes.
6. Clean condenser coils (use care – salt water has made fins brittle).

Savings   Normal Load   180 tons

KW Reduction Average 1.25 KW/ton to 1.02 KW/ton
Estimate: 5,000 HRS x .23 KW x 180 x .08/KW – $16,560

   3000 HRS at 1/2 Capacity – $4,968

Total $21,528/year

Figure 13-1 Roof Top Air Cooled Chillers. Wind blows discharge air back into intake.

COUNTRY CLUB

Dining room cooled 25 hours per day at 68deg to get 75deg at noon. AC unit 3 fan plugged, cleaned up to get 6800 cfm.

Conditioning Air Supply Test

* – Low Air Supply

AC Unit	Test Cubic Feet/Min.	Specification From Drawings
1	9952	9450
2	3192*	4300
3	800*	7200
4	2400*	3200
5	3420	N/A
6	800	800
7	5800*	7000
8	800	800
9	2480	2800
10	1480*	4000
11	1666*	4000
12	2331	2000

Room Temperature Test

June 26, 1992, 11:00 A.M. to 12:30 P.M.
Outside Air 91° – 63% RH)

Dining Room – East Window Section	73°	–	73% RH
West Window Section	74°	–	65% RH
Dance Floor	76°	–	67% RH
East Central	75°	–	70% RH
Northeast Section	74°	–	90% RH
Northwest Section	75°	–	90% RH
Bar	75°	–	73% RH
Cocktail Lounge	75°	–	70% RH
Lobby	75°	–	73% RH
Ladies Card Room – South	72°	–	70% RH
North	73°	–	77% RH
Office	73°	–	78% RH
Men's Card Room – South	75°	–	70% RH
North	76°	–	67% RH
Men's Locker Room	75°	–	65% RH
Pro–Shop	72°	–	75% RH

Variable Air Volume System

Bypass duct at unit to control static pressure when zone outlets close on reduced cooling or heating.

Ducts too small for maximum bypass.

Test Results
A/C Unit #1 – 6000 CFM

Area	Specified CFM	Actual CFM	Thermostat Setting	k° Indication	Room °F Temperature
Kitchen	900	0	74	82	76
Banquet	2800	924	68	78	74
Offices	875	327	72	72	70
Media-Foyer	1200	647	68	76	76
Switch Gear		200			
	5775	2098			

Outside Air 1760 CFM at 89°

A/C Unit #2 – 5000 CFM

Area	Specified CFM	Actual CFM	Thermostat Setting	k° Indication	Room °F Temperature
Fitness	780	785	62	72	69
Aerobics	800	632	68	73	72
Corridor	800	391	70	78	74
Foyer					
Womens Card					
Locker	1000	467	70	75	73
Mens Card					
Locker	1000	611	69	78	72
Billiards	600	268	68	85	74
	4980	3154			

Outside Air 1090 CFM at 89°

Bypass Duct Capacity

A/C Unit #1 – 5775 CFM
Bypass 2 – 14" ducts @ maximum of 1890 CFM each.
Maximum bypass 5775 CFM – 875 CFM – 4900 CFM.
Maximum capacity of bypass – 2 x 1890 – 3780 CFM.

A/C Unit #2 – 4980 CFM
Bypass 2 – 12" ducts @ maximum of 1260 CFM each.
Maximum bypass 4890 – 600 – 4290 CFM.
Maximum capacity of bypass – 2 x 1260 – 2520 CFM.

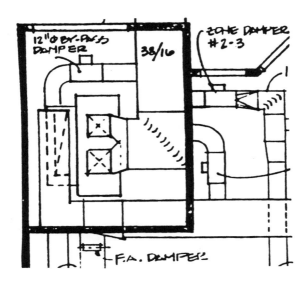

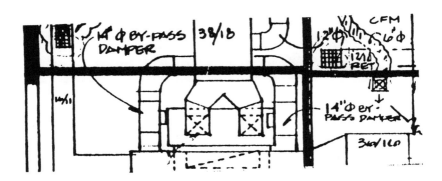

Figure 13-2 Variable Air Volume System. Bypass duct at unit to control static pressure
when zone outlets close on reduced cooling or heating.
Ducts too small for maximum bypass

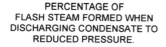

PERCENTAGE OF
FLASH STEAM FORMED WHEN
DISCHARGING CONDENSATE TO
REDUCED PRESSURE.

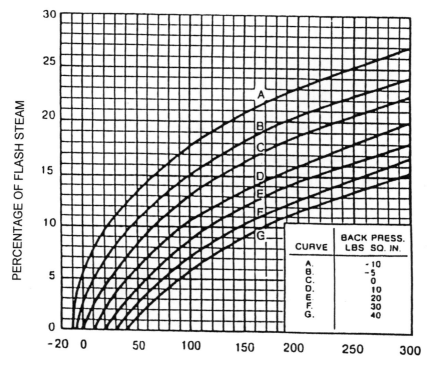

Figure 13-3 PSI from which condensate is discharged. Flash steam losses at condensate return temperatures above 225deg. See page 189

OFFICE BUILDING
20 stories, 18 years old, 300,000 square feet, location in Midwest.

Boiler room on roof.

Four boilers, gravity gas, cast iron, 10,500,000 BTU/HR each, 15 PSI steam.

Cooling, 1,400 tons absorption chillers. Central air supply to induction units – perimeter, boiler room under negative pressure. Exhaust fan in boiler room to reduce heat.

Gas flame 12" height

Flue gas test 15% carbon monoxide, heating provided by steam-water heat exchanger on chilled water line.

In summer, thermometer on chilled water supply to core units – 82°

Energy use nearly 3 times normal.

Negative pressure through lobby entrance doors.

What is wrong?

What is necessary to correct?

Solution

1. Reversed rotation of exhaust fan to provide combustion air to boiler room by changing two wires on motor.
2. Correct leaking steam valve on heat exchanger on chilled water supply. Exhaust fans in elevator penthouse pulling air from shaft. Recommended outside air louver in machine room.

Results:

Boiler combustion efficiency raised from 45% to 80%.

Chilled water reduced to 45° to core, central air supply units from 82°.

Savings from boiler operation alone $200,000/year.

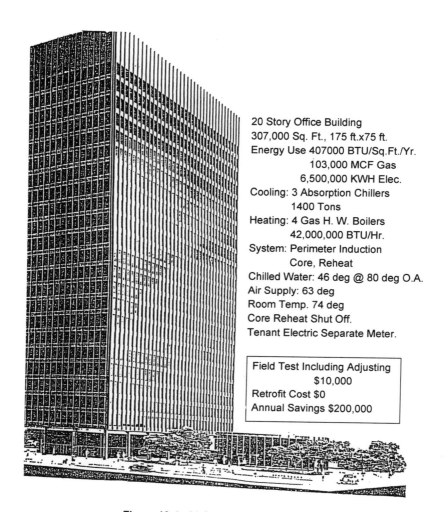

20 Story Office Building
307,000 Sq. Ft., 175 ft.x75 ft.
Energy Use 407000 BTU/Sq.Ft./Yr.
          103,000 MCF Gas
          6,500,000 KWH Elec.
Cooling: 3 Absorption Chillers
          1400 Tons
Heating: 4 Gas H. W. Boilers
          42,000,000 BTU/Hr.
System: Perimeter Induction
          Core, Reheat
Chilled Water: 46 deg @ 80 deg O.A.
Air Supply: 63 deg
Room Temp. 74 deg
Core Reheat Shut Off.
Tenant Electric Separate Meter.

Field Test Including Adjusting
          $10,000
Retrofit Cost $0
Annual Savings $200,000

Figure 13-4 20 Story Office Building

OFFICE BUILDING

77,000 Sq. Ft., 10 Stories, South Florida, 4 Years Old, 80% Occupied

Air Conditioning:

– Individual heat pumps.

– 270 ton cooling tower.

– 810 GPM Condenser water system.

– 66° Central outside air supply.

– Central Management System.

– 8,500 CFM toilet exhaust.

– 1989 Electric Bill – $132,000

– Demand 336KW – 444KW

Field Test Indicated Corrections:

		Savings/Yr.	Cost
1.	Reduce outside air from 17,000 CFM to 8,500 CFM.	$ 5,000	$ 0
2.	Air Condition Elevator Penthouse.	1,600	0
3.	Reduce night and weekend condenser water circulation	6,000	2,500
4.	Demand reduction program.	10,000	0
		$22,600	$2,500
	Engineering Test Cost:		$3,000

HOSPITAL

Patient's Rooms – System – Multizone Reheat.

Basic Unit

A. Supply – Axial Flow Fan – Fileter – Heat Wheel – Cooling Coil (Chilled Water) – HEPA Filter – Supply Duct.
100% O.A. from Equipment Room.

B. Exhaust – Duct – Heat Wheel – Axial Flow Fan – Duct – Outside.
100% Exhaust to Outside.

C. Control – Constant supply temperature. Individual Room – Reheat hot water coil 8" x12", 1 row, 140° water, thermostat – pneumatic, hot water valve – pneumatic.

D. Individual Room and Supply Air temperatures measured. On return from full heating to room temperature thermostat setting, reheat coil took 20 minutes to cool off.

E. Unit tested. Considerable leakage of air around heat wheel.

Solution:

What is wrong with system?

1. Bad thermostats.

2. Bad valves.

3. Hot water temperature too high.

4. Heat wheel has high static pressure and leakage.

5. Heat wheel leakage reduces air supply.

How to correct.

1. Replace bad thermostats.

2. Replace bad valves.

3. Reduce hot water temperature to 90°.

4. Replace heat wheel or change system to partial recirculation.

ROOM TEMPERATURE READINGS

Room No.	Room Temp.	Stat Setting	Air Supply Temp	Maximum Air Temperature		Defective		Comments
				Cool	Heat	Stat	Valve	
Day	74	61	67	65.5	95.5	X	X	Supply Air Temp. .56°
529	75	71	75.5	64	104.5		X	Outside Air 73°
528	73.5	71	58.5	58.5	93.5			
527	74.5	84	105	69	105	X	X	
526	76.5	85	95.5	68	95.5	?	X	
525	74.5	82	90.5	66	90.5	X	X	
524	74.5	84	92	74	92	X	X	
523	73	75	97	67	97		X	
522	72	58	58	58	91.5	X	X	
521	77	73	78.5	66	90	X	X	
520	76	77	74	57.5	74	X	X	
519	79.5	85	100	68	100	X	X	
518	76.5	77	94.5	60	94.5		X	
517	76.5	86	97.5	63.5	97.5		X	
516	76.5	71	57	57	95.5	X		
515	78.5	80	106	60	106	X	X	
514	78.5	73	107.5	63	107.5	X	X	
513	77	77	79.5	60	90		X	
512	ISOLATION							
511	77.5	92	91.5	67	92	X	X	
510	76.5	83	92	71	92	X	X	Grill Closed
509	77	74	92	74	92	X	X	
508	77.5	79	84	64.5	85		X	
507	77.5	72	98.5	70	98.5	X	X	
506	77	75	85	62.5	98.5	X	X	
505	77	77	98	75	98	X	X	
504	75.5	77	93.5	67	101	X	X	
503	76.5	76	72	61	104.5		X	
502	76	68	66	62.5	102.5	X	X	
SW Cor.	76.5	84	60	56.5	56.5	X	X	
NW Cor.	76					?	?	No Stat
SE Cor.	77	60	59.5	58.5	58.5	X	X	
NE Cor.	77	56	57	56	68	X	X	

HOSPITAL

A/C Unit	Specified	1970 Test	1977 Test	1983 Test
Condition				
Supply Air CFM	20,600	16.470	11,863	13,644
Suction Static Press In. H20		−.36		0
Discharge " " " "		+3.50		+3.5
Total " " " "	3.5	3.86	2.4	+3.5
Supply Duct " " "			.61	.50
Fan Motor HP	20	20	20	20
" " AMPS	25	20.2	22/22/22	20.5/20.5/20
" " Volts	480	460	500	500
Fan RPM	797	815	865	879
Exhaust Air CFM	20,600	17,650	11,837	11,797
Suction Static Press In. H20		−1.75		−1.3
Discharge " " " "		+1.0		
Total " " " "	2.0	2.75	2.5	
Fan Motor HP	15.0	15.0	15.0	15
" " AMPS	20.0	20.2	15.7/15.8	18.8/18/18
" " Volts	480	460	500	500
Fan RPM	718	815	705	705?
Heat Wheel				
Supply Fan Ent.Air DB	73.5	81.8		64
Ent.Air WB	69.2	69.0	61	
Lvg.Air DB	74.6	79.5		65
Lvg.Air WB	64.0	66.8		60
Exhaust Fan Ent.Air DB	75	77.8		
Ent.Air WB	62.3	65.0		
Lvg.Air DB		80.5		
Lvg.Air WB		68.0		
Cooling Coil Ent.Air DB/WB	79.1/67.6	79.5/66.8		
Lvg.Air DB/WB	56/55	57.5/56.8		56/52
Chilled Water Ent.	50°	51.0		
Lvg.		57.5		

HOSPITAL
Leaking Heat on Cooling Cycle

45h Floor– 62°	5th Floor–56°
°F High 4	9.5
4	8
10	2.5
0	13
6	12
8.5	10
5.5	18
5	11
5.5	2
5	10
1.5	1.5
9.5	12
9	4
2.5	7.5
5	1
1.5	4
2.5	7
1	4
2	11
2.5	15
19	18
1.5	8.5
3	14
2	6.5
2	19
6	11
0	5
0	6.5
2	.5
2.5	2.5
3	0
3	
2	

$33) \overline{136} = 4.2°$ $31) \overline{255} = 8.3$

Average $\dfrac{4.2° + 8.3°}{2} = \dfrac{12.5}{2} = 6.3°$

Estimated Savings after Correction

Required CFM 18,350
Electric Cost for Overcooling

$$\frac{18350CFM \times 6.3° \times 1.08}{12,000} = 10.4KW$$

10.4KW x 8760Hrs x .05¢ = 4,555.00

Fuel Oil Cost Due to Overheating

$$\frac{18350 \times 6.3 \times 1.08}{132,000 \times .75} = 1.33 \text{ Gal/Hr.}$$

1.33x8760x.95¢ = $11,068.00
Total Cost   4,555.00
                   $15,623.00 Total Annual
                               Savings with
                               Corrected
                               Controls

DEPARTMENT STORE

Manual operation, start up 6am., shut down 9pm.

Program step start up on motors. Electric utility recorded demand every 15 minutes for one year. Examination of records indicated that system was not shut down 108 days in one year. Cost – $75,000 per year

Time	06/22	Time	06/22	Time	10/23	Time	10/23
0000	92	15	551	0000	278	15	492
15	92	30	550	15	264	30	496
30	93	45	549	30	243	45	494
45	93	1300	551	45	234	1300	493
0100	92	15	554	0100	234	15	493
15	92	30	556	15	234	30	493
30	93	45	555	30	233	30	493
45	93	1400	547	45	233	1400	494
0200	91	15	541	0200	233	15	496
15	91	30	542	15	232	30	497
30	93	45	543	30	232	45	497
45	92	1500	546	45	232	1500	497
0300	92	15	547	0300	232	15	495
15	93	30	546	15	232	30	495
30	93	45	543	30	232	45	493
45	92	1600	542	45	232	1600	490
8400	91	15	542	8400	232	15	489
15	92	30	537	15	232	30	488
30	92	45	535	30	231	45	485
45	93	1700	537	45	232	1700	483
8500	94	15	535	8500	233	15	481
15	98	30	533	15	235	30	482
30	119	45	533	30	239	45	481
45	228	1800	531	45	260	1800	476
0600	325	15	531	0600	323	15	474
15	375	30	529	15	378	30	472
30	460	45	533	30	387	45	476
45	495	1900	544	45	381	1900	483
0700	488	15	553	0700	378	15	488
15	484	30	563	15	380	30	496
30	477	45	570	30	384	45	505
45	469	2000	565	45	385	2000	508
0800	472	15	561	0800	385	15	504
15	480	30	556	15	390	30	496
30	485	45	525	30	402	45	484
45	494	2100	394	45	414	2100	458
0900	522	15	225	0900	433	15	408
15	544	30	138	15	458	30	330
30	549	45	109	30	471	45	282
45	552	2200	105	45	476	2200	277
1000	551	15	98	1000	477	15	266
30	550	45	84	30	478	45	252
45	554	2300	81	45	480	2300	249
1100	557	15	81	1100	481	15	249
30	553	45	80	30	490	45	248
45	553			45	489		

RESIDENTIAL

Chiller Test

46.5° LWT, 50° EWT operation
Ambient air temperature 78°
68.2 amps at 208 volts

Discharge fan readings:

SE fan	3720 CFM at 96°	
NE fan	3220 CFM at 77°	
NW fan	4200 CFM at 76.5°	
SW fan	3460 CFM at 95.5°	
	14600 CFM total	

Net compressor capacity:

3720 CFM X (96° – 78°) x 1.08   = 72,316 BTU/Hr
3460 CFM X (95.5° – 78°) X 1.08   = 65,394
                                  137,710 BTU/Hr

Electric compressor load, assuming .85 power factor:

68.2A X 208V X 1.73 X 3.412 X .85 = 71,173 BTU/Hr

Net capacity: 66,536 BTU/Hr = 5.5 tons.

Compressor KW:

68.2A X 208V X 1.73/1000 X .85 = 20.86 KW

rated capacity at 46° LWT and 85° condensing temperature = 39.6 KW, 39.5 tons cooling, 94.7 GPM.

Net chiller cold water production should be:

(20.86 KW/39.6KW) X 39.5 tons = 20.81 tons

Compressor Unloaded Stuck Open.

Chiller is producing only 5.5 tons.

At rated capacity:

(20.81 tons X 12000)/(500 X 61 GPM) = 8.2° water temperature
                                      difference between in and out.

Chilled water should be leaving chiller at $50° - 8.2° = 41.8°$ instead of $46.5°$.

Total condenser fan capacity:

measured	14,600 CFM
catalog rated	26,000 CFM

Chiller rating   39.6 KW/39.5 tons = 1 KW/ton

Measured   20.86 KW/5.5 tons = 3.79 KW/ton

Chapter 14

Air Pollution

Air pollution or "Sick Building Syndrome" has become a major part of the operation of air conditioning and heating. The data you have obtained and are now analyzing can help solve the pollution problems.

(1) *Temperature and humidity* – Most people are comfortable in the effective temperature range as determined by ASHRAE tests. Since about 40% of the heat produced by our bodies is in evaporation, the air surrounding us must move so that less humid air can absorb the moisture.

Fifty feet per minute velocity is considered still air, so more air motion is needed. By using the air flow meter, this velocity can be determined in the space around a person's body. A minimum of 100 feet per minute or roughly one mile per hour is necessary.

If the humidity is high, more air motion is needed than if the air is dry. A good range of operation is 30 to 70%.

This is the reason that the air patterns and the air distribution described in pages 210 and 211 are necessary as this affects the comfort of personnel. Many of the problems of air distribution over the bodies is due to a change in the air velocity from a designed air outlet, especially in the ceiling or walls, or from partitions or other materials that may interfere with the distribution of air. Also, since the body loses the rest of the heat by radiation, if there is interference of this due to sun shining on our bodies, or standing or working near a warm machine, this also causes discomfort and requires cool air to be distributed over the body.

(2) *Experience* – An 8,000 square foot office space had about 20 people in it and two of the women complained of dry throats and coughing. The tenant was about to move out when the Owner asked us to run a test on the air quality. We ran chemical tests for formaldehyde, ozone, carbon monoxide, carbon dioxide, and could not find anything. We then took out our air flow meter and went around the space and discovered that there was practically no air motion over anybody's body and that the perforated ceiling outlets were dumping air straight down to the floor and the air within two feet of the floor was 3 degrees colder than the rest of the space. We also measured the temperature and humidity at 72° and 37% which is unusually low for a southern climate. Further tests indicated that the air quantity coming out of the ceiling outlets were much less than the original design so that it was readjusted to get better air circulation and the problem disappeared. The system also was heat pump units mounted in the ceiling which were examined and found that there was about 1/2-inch of water laying in the drain pan because of the location of the drain connection. The drain pan was full of slime which was also adding to the bacteria content in the air. The drain pan connection was changed from the side to the bottom which contributed to the solving of the problem.

Many people require different temperatures due to physical problems or activity. For instance, a person seated at work gives off about 400 BTU per hour, and a person very active or working at machinery can give off as much as 1,700 BTU per hour, thus requiring cooler air for those people who are active than those people who are seated with minimum activity.

If a person is sitting near a window where the sun may be shining, and if he does not pull down shades, venetian blinds, or use some form of interference with the sun's direct rays, he will feel much more heat and may require considerably more cooling over that part of the body which is exposed. We have tested solar glass windows on the west side at 2:00 PM with an outside temperature of 80° and the glass surface temperature was 145°. Since the window was not covered with any form of shade, the air temperature within the space within three feet of that window was two degrees warmer than the rest of the space.

(3) *Chemical Contaminates* – Many of the materials in a space such as carpeting and furniture may give off certain chemicals called "off gases" such as formaldehyde and ozone. Many of these chemicals have now been taken care of with new materials, but some machinery still goes off certain chemicals. For instance, a copy machine can give off ozone and if in strong enough quantities, and if a person stands there for a very long period of time, may be affected with respiratory problems. In those cases, the air motion should be measured and, if possible, the air should be returned to the system from that particular location directly so that the concentration of ozone is reduced. Carbon monoxide may occur in the air if the ventilating system is not properly designed and may pull air from a parking garage below and draw it into the system.

The location of the outside air intake is extremely important, especially if it is located near a toilet vent, cooling tower discharge, or whether a boiler flue or incinerator flue may be located so that wind blowing the right direction puts these contaminates into the outside air intake. In incinerators especially, there may be certain chemicals like phosgene, cyanates from urethanes being burned. In boiler flue gasses, there is carbon monoxide and if freon refrigerants are present in the air, freon breaks down to phosgene at about 1200° and at 2400° breaks down to carbon dioxide and hydrochloric acid.

Some type of dust particles can produce allergic conditions so that the filters should be in very good condition and should remove most of the contaminates in the air, especially materials like fiberglass, asbestos, lint, and deteriorated paper products.

If chemical materials are used for cleaning purposes, or for other activities in a space, the cleaning materials in the atmosphere should be removed before the space is occupied so that if you are using certain chemicals for cleaning floors or walls, make sure that the air system is flushed out before people occupy it the next day.

(4) *Bacteria and Organic Materials* – In dry climates, moisture has to be added to the system to get it up to at least 30% so that the method of supplying this moisture should be such that bacteria or any organic materials does not grow either in the drain pan of the air conditioning unit or in the duct work. In humid climates, a major problem is mold and mildew which seems to grow on almost

all surfaces, and also algae and similar materials in drain pans. The dewpoint of the air is a major factor in what happens to the generation of these bacterial materials since most of them require moisture to multiply and to increase in quantity. One of the major problems of most systems is the drain pan which, if it doesn't drain properly, can collect algae and other material growth, some as serious as Legionnaire's Disease. In order to determine what happens in the drain pan, the method is to open up the unit and feel the surface and if the drain pan surface is slippery, you're generating an algae growth. To eliminate this problem in drain pans, certain materials like biocides are used. If this equipment or material is used, care should be taken that it does not interfere with the air supply condition, as some of these materials are so strong that they can get into the air stream and cause more contamination and problems to the personnel. One of the major problems with mold generation is when air conditioning systems are shut down at night or on weekends,and no control is provided to keep the humidity from rising, especially in humid climates where the average room temperature is about 72° which means that all the materials in that room are at 72°. Quite often the outside air at night time can have a dewpoint as high as 80° which means that if this air gets into the building, all the surfaces in the space get wet; and when they get wet, mold grows on it and the next morning when the system starts up, all this mold is distributed throughout the space and if it gets concentrated enough, it causes respiratory problems. The solution for this problem is to seal the building up at night time by putting motorized dampers on the outside air intakes that shut when the system stops and use at least back-draft dampers on all exhaust fans through the roof and some means of closure of all other openings in the building. This has solved a number of problems, some of which were quite serious. Not only does this occur in the southern climates, but can occur in the upper mid-west, especially in the summer, and especially when it rains. If the system is shut down more than 1/2 a day, it is recommended that the system be turned on at full capacity for at least two hours a day to remove the moisture which may be generated and to prevent the generation of mold and mildew in the space. If the building is only partly air conditioned, the conditioned space should be sealed off from the balance of the building so the moisture does not penetrate to that space.

(5) *Outside air supply* – Although the code recommendations require approximately 20 CFM of outside air per person, increasing the amount of outside air to eliminate air pollution may not always do the job since the outside air may be more contaminated than the inside of the building. So, it is preferable to clean the interior air and re-use it wherever possible except in those cases where exhaust is required such as in hospitals and kitchens where the contaminated air should not be recirculated. One of the problems with heat recovery devices, especially in hospitals, is if the exhaust air is not completely free of contaminates such as from isolating rooms or toilets, that these contaminates can be recirculated back into the building through those so-called heat wheels.

(6) *Supply Air Temperature* – Although there is a program to reduce air quantities and reduce the supply air temperatures, this may be alright in low humidity areas, but in high humidity areas, this causes a number of problems. If the supply temperature is much below 55°, there is a tendency to grow mold on the air outlets as well as inside the ducts due to the high dewpoint of the air surrounding them. This also increases the amount of mold generation at night time, when the system is shut down because of the low temperature of the surfaces in the duct work and the air outlets.

Some of the causes of low supply air temperature is dirty filters, dirty coils, and especially dirty fans of the forward curve blade type. That little curved blade will reduce capacity considerably if it gets partly filled up with dirt.

(7) *Experience* – A country-club house had a large dining room. They ran their air conditioning system all night at 68° in order to get 75° at lunch time. The drawings were checked and it was discovered that the air units should have handled 8,000 CFM. Test was actually made on the unit and it was only handling 800 CFM. After examining the unit, it was found that the fan blades were partially filled with dirt. There was a little bit of dirt on the coils. This was cleaned up, the unit brought up to 6,800 CFM. The system now is turned on one hour before lunch time at 74°.

PROCEDURES FOR ANALYZING A BUILDING TO SOLVE "SICK BUILDING SYNDROME"

(1) Find out if you really have a pollution problem.

     A. If building occupants question condition but are not actually sick or uncomfortable, reassure them by testing.

         1. Measure temperature and humidity in each 100 square foot area.

         2. Determine if there is air motion in all areas, especially at breathing level of 5-feet above floor.

         3. Measure air temperature of supply at air outlet and at air conditioner.

         4. Measure temperature at inside of window glass east at 10:00 AM, south at 12:00 Noon, and west at 4:00 PM. If sun is shining on glass, tell occupants to use shades or blinds when the sun is shining.

         5. Measure temperature above ceiling if removable and used for return air. Inspect for asbestos. If space above the ceiling is used for return air, look for dust, dirt, and asbestos. Temperatures lower than return air indicates duct leak or VAV control bypass units. If asbestos is present and return air can be brought back to air conditioning unit below ceiling, asbestos may not have to be removed. If dusty and dirty, check filters and air conditioning unit for efficiency.

         6. Inspect inside of air conditioning unit for slime, mildew, and algae in drain pan and coils. Since most bacteria carried in an air conditioning system occur in wet drain pans and on cooling coils, most drain pans are equipped with a trap which almost always plugs up with algae. Make provisions for cleaning coil drain pans and drain line often, at least once a month on drains and once a year on coils. If possible, provide plugs removable on the trap from the drain pan so that the trap can be cleaned out easily.

         7. Inspect outside air intake location for admission of pollutants, cooling tower discharge, auto exhaust, incinerator exhaust, toilet and building exhaust, and toilet plumbing vents. Since outside air intakes may bring in contaminated

water spray from cooling towers, carbon monoxide, and hydrogen dioxide from autos, chemicals from incinerators, and contaminated air from toilet vents and exhaust, these should be checked very carefully.

8. Examine air supply outlets for black specks of mildew.

9. If air conditioning system is shut off nights and weekends, make sure there is motorized dampers on the outside air intakes that will close when the fans stop and make sure that there are self-closing dampers on all openings in the building such as supply and exhaust openings and fans.

10. Measure outside air entering or leaving building through doors for air velocity.

11. Find out what cleaning chemicals are used by janitorial service and where these chemicals are stored. Some chemicals require complete clear out of air in the building, after use.

12. Measure outside air supply to air conditioning units for sufficient quantity for ventilation.

13. Measure toilet exhaust air quantity and determine when exhaust fans run.

14. Examine interior of outside walls and ceilings for appearance or odor of mold. There is a possibility that moisture might be leaking through the walls into the insulation in the wall space.

15. Examine thermostat for temperature setting indication and actual room temperature. Fan setting should be in the "on" position for continuous air flow in commercial buildings.

16. Determine if new carpeting, wall covering, or furnishings have been installed in the past month and what type of adhesives or materials are in them in case they are giving off what we call "off gases". If the air conditioning system is a VAV, variable volume type, is there sufficient air circulation at low air volumes.

17. If there is a kitchen or laundry, check to see if there is a negative air pressure preventing odors and lint entering other occupied space.

18. If there is an exposed water such as a fountain or waterfall, determine if the water is clean.

19. If the building has a smoke pressurizing system, are there dampers on the air intakes which are necessary to seal the building at night time.

20. If a heat recovery device is used on the air conditioner, inspect to see if it is clean and does not recirculate exhaust air into the supply system.

Index